Robert Stirling's Models of the "Air Engine"

by James G. Rizzo

Camden Miniature Steam Services.

British Library Cataloguing-in-Publication-Data: a catalogue record of this book is held by the British Library.

First Printing 2009

ISBN No. 978-0-9547131-9-5

Also by the Author:

AN INTRODUCTION TO THE STIRLING ENGINE – Malta University Press, 1982

MODEL STIRLING AND HOT AIR ENGINES – Malta University Press, 1983

MODELLING STIRLING AND HOT AIR ENGINES – Patrick Stephens, 1985

THE STIRLING ENGINE MANUAL (Volume 1) – Camden Miniature Steam Services – 1995, 1997, 2000, 2005)

THE STIRLING ENGINE MANUAL (Volume 2) – Camden Miniature Steam Services, 1999

Published in Great Britain by:

CAMDEN MINIATURE STEAM SERVICES
Barrow Farm, Rode, Frome, Somerset. BA11 6PS
www.camdenmin.co.uk

Layout and design by Camden Studios.

Printed by
Salisbury Printing Co. Ltd.

Front Cover: The models shown are ½ scale replicas of the 'Air Engines' originally made by the Rev. Robert Stirling D.D. and donated by him to the University of Edinburgh and the University of Glasgow, Scotland.

IMPORTANT:
Some of the processes and working methods described in this book can potentially be hazardous. The author and publisher are just passing on information - your safety, and that of others, is entirely your responsibility.

Table of Contents

The Rev. Robert Stirling's Models of the 'Air Engine'

The models of an air engine made by Robert Stirling himself are the more historical because they are an everlasting legacy of a great man with tremendous foresight.

It was only fitting that the Rev. Robert Stirling presented a model to each of the two Universities where he studied, the University of Edinburgh and the University of Glasgow, Scotland.

Rev. Robert Stirling

Doctor of Divinity, Minister and Inventor,
whose patents were the basis on which modern Stirling engines are developed, and whose name has been coined to describe this genre of engines.

ACKNOWLEDGEMENTS

This book was made possible with the generous help of a number of persons, mostly from Scotland.

Foremost of the persons who helped towards the fruition of this project is Dr. Marie Laurie, whom we had the pleasure to welcome, during her visit to Malta, in our holiday home in St. Paul's Bay in September 2004. In the course of conversation somehow my hobby cropped up and my most recent publications shown. As an erudite Scots woman the connection between a Maltese hobbyist and Scotland's famous son, the Rev. Robert Stirling, D.D., must have been somewhat of a cultural shock. Marie gave me her whole hearted support when I asked for her help in making contact with the Royal Museum of Edinburgh and the University of Glasgow for photos and information on both Stirling's models of the air engines, as they were called in his era. Moreover Marie kept in touch with her contacts and furthermore undertook research into archived documents. I sincerely doubt whether I would have been able to continue the project without Marie's help.

The first to respond and with great fervour was Alan Mills of the National Museums of Scotland. I thank him most heartily for the wealth of details he sent me. Following in his footsteps Ian Gentle, of N.M.S., not only spared no effort to fill in some gaps, point me in the right direction but also put up with my many requests for clarifications. To Ian, another big 'thank you'.

Photo 1: Dr. Marie Laurie and Ian Gentle of NMS with Robert Stirling's model of the 'air engine' now at the Royal Museum, Edinburgh (Courtesy - Alan Mills)

The Hunterian Museum of the University of Glasgow houses the second of the Rev. Stirling's models which he donated around 1826. Mr. Mike Jewkes, of the Hunterian Museum took great patience to measure, record and describe many of the components of this model. To Mike, many thanks for your valuable assistance.

To these four persons, my full gratitude – their great co-operation makes me more appreciative of Scottish hospitality and friendly courtesy.

I am grateful to the Parish of Galston, Ayrshire, Scotland, for helping me to identify areas closely associated with the Rev. Robert Stirling, particularly the Old Manse where he lived for 54 years and the Cemetery where he was buried. Robert Murray, Sessions Clerk, and Ian Thomson, Beadle of the Parish Church, went out of their way to make sure that I had the right information. John Connell managed to trace the photo/postcard of the Old Manse in the possession of Henry Kerr, former Sessions Clerk, thus ensuring, through Robert Murray, that this historical building, now demolished to make way for new development, is recorded for posterity.

Roy Darlington helped with his advice on leather pistons which I had not used before, and moreover gave me permission to use two images from his excellent book (together with co-author Keith Strong) 'Stirling and Hot Air Engines'. Keith Strong, on his part, a CAD and CAE specialist, as well as a design engineer (automotive and astrospace), who specialises in Stirling technology, kindly accepted to draw and render the CAD components that would normally be difficult to interpret from line drawings.

Other persons who have been instrumental in making this book possible include Cyril Dennis who kindly took photos of his pristine Philips Stirling generator and gave me permission to use these. Raymond Calleja, a fellow member of the Association of Model Engineers (Malta) helped me with some delicate machining of the models' columns and taught me how to make jigs for repeat copies. Lorrie Saliba, an old colleague and friend from my Scouting days came to my rescue when I could not find the right type of wood for the columns – not only the right wood, but the right size ready for machining. Pierre Bundy dealt with many of my photography problems, before the digital age and more so afterwards, and with his technical ability and admirable patience solved almost all my problems.

Finally to my three sons, Adrian, Alex and Stephen who all contributed a great, great deal with their I.T. knowledge, equipment and time – 'thank you, boys'.

James G. Rizzo
Iklin, Gozo and Qawra (St Paul's Bay)
MALTA
2005 – 2007

THIS BOOK IS DEDICATED TO:

JUSTIN, LUKE, RYAN and SHAUN, grandsons,
(in alphabetical and chronological order)

Introduction

One can never be sure whether the models of an air engine made by the Rev. Robert Stirling were meant to be used for demonstration purposes or as a memento of his years of learning at the two Universities, or both. But the fact that both models did run and were actually used for demonstration purposes albeit many years later proves that they are a very important historical record as well as legacy of a far-seeing engineer who believed in a new way to make a safe prime mover that could compete with the steam engines of that era. These models were meant demonstrate his beliefs.

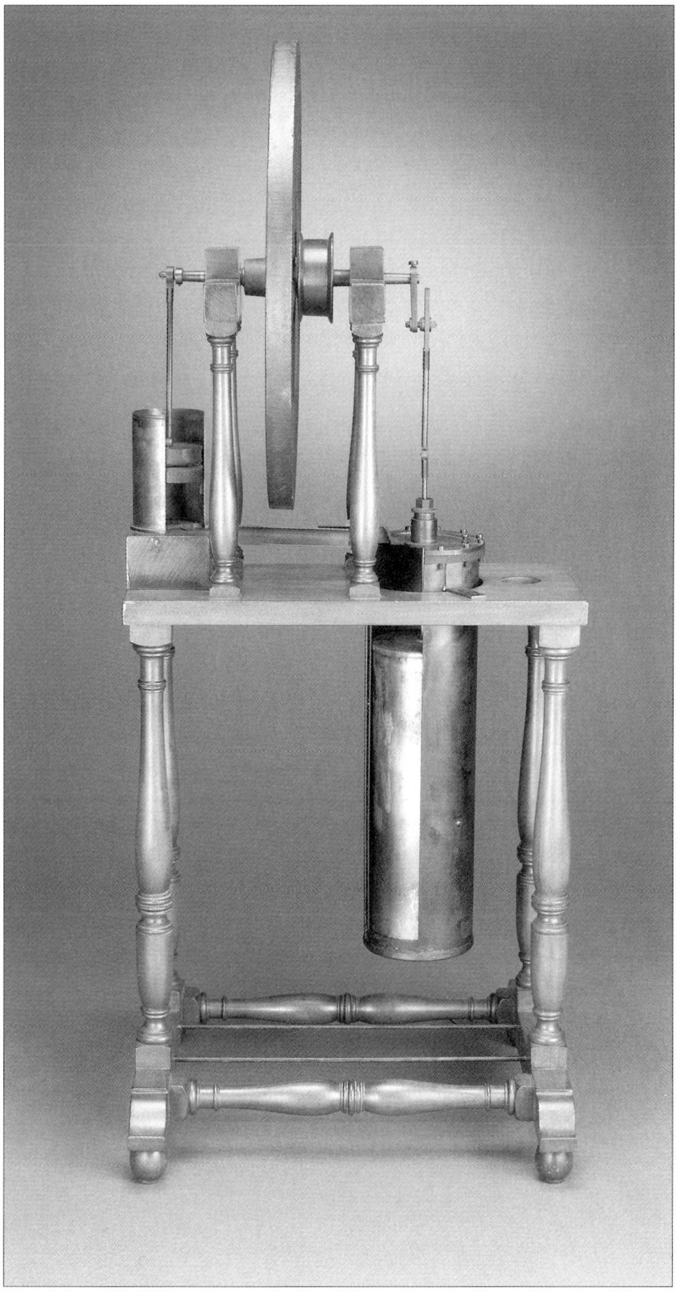

Photo 2: Cut-away replica of Stirling's Edinburgh model at the Science Museum, Kensington showing details of the displacer and the power piston (Courtesy - Science Museum)

The first time I saw a replica of the Edinburgh model was in 1983 during my first visit to the Science Museum in Kensington, London, (Photo 2). Although at the time I was just a beginner to the hobby of Stirling engineering, I felt that one day I had to construct a replica myself in honour of the great achiever. Lack of specifications, inability to obtain additional information, close-up photos and to measure the engine precluded me from carrying out this project for twenty-two years. In the meantime I took up other areas of this hobby with some degree of success.

Some years later I came across a photocopy of an article by J.M.A. Lenihan and Johnston R. McKay entitled "Dr. Stirling and his air engine", which I filed away with a huge amount of technical papers. My copy of the article had a faint image of another model made by the Robert Stirling, this one held in the University of Glasgow. The photo showed a similar model to the Edinburgh one but with one very visible addition, a water cooler.

I could have constructed a version of the Edinburgh engine from images in various publications, an old colour photograph I took in 1983 and with the help of Robert Sier's line drawing of the drive mechanism, but it would not have been accurate. With additional images, close-up and full view, and specifications obtained from the N.M.S. I had enough details to make a reasonably accurate replica of at least the Edinburgh model to start with.

A decision had to be made whether to reproduce a full-scale replica or a scaled-down version. Both original models are 29 inches (737 cm) high. These would have been a bit too large, both for my collection of vintage engines as well as to carry them about for demonstration. So a half-size, table-top version was decided on, at the same time keeping within the range of other vintage models I have built, such as a Rider-Ericsson Pumping engine, a Bremen Caloric Engine, and Anderson/Eriksson caloric engine and two Van Rennes hot air pumping engines.

Robert Stirling's models are elegant to look at and this appearance had to be maintained in scaling down the replica. It was felt that a 50% scale model, if accurately constructed, would have the same visual impact and moreover would perform just as well. Admittedly scaling up or down a hot air engine does not guarantee the same efficiency – this has been proven many times over. These models were never meant for high revolutions or for power testing. It can be visualised that the original models when used for demonstration were run at a low speed – any higher speed would have shaken the wooden structure with such a high engine profile and layout. The frame work is too frail for such tests.

The present state of the Rev. Stirling's models

Over 180 years have passed since the models were made and that is a long time for any mechanical device. Considering that the frame of these models was made from wood, which is liable to warping with dampness and shrinkage due to changes of temperature, that the frame is fairly flimsy and that glue was used to join the columns to the engine platform and base, it is a wonder that these models have survived the test of time with minor deterioration.

The model held in the Royal Museum of Edinburgh is perhaps in better shape, probably partly because the model was better preserved and also because it was rarely run for demonstration purposes. In 1975 the model was presented by the University of Edinburgh to the Royal Museum of Edinburgh for preservation and display. Having seen the model quite close-up I can confirm that it is in pretty good shape.

The model held in the Hunterian Museum of the University of Glasgow has not fared as well. For about 20 years the model was placed in storage and forgotten until discovered by Prof. William Thomson in 1841. Prof. Thomson (later Lord Kelvin) had the model dismantled, rehabilitated with some changes in components, and got it running. From information received by the author it appears that other parts were replaced at later times. Particular mention was made of the displacer rod which works loosely in the gland as 'a more recent replacement and is of modern mild steel'. On the other hand there are documented records that show that this engine was used in lecture room demonstrations as late as 1876.

Having acknowledged the great help I have received, and having made a thorough study of the two engines and all the components as well as having seen both models at close quarters I feel that the details given in the following chapters are as close as possible to the actual dimensions. I have seen the Royal Museum's model very closely and from all angles and the details given to me appear to be correct to the closest millimeter.

Although I have also seen the Hunterian Museum's model it has not been possible for me to get too close as it is enclosed in a sealed show case with only a front glass or Perspex window. It was not possible to look at the model from either side. I had therefore to rely on specifications given to me by the Museum Staff. These specifications, with minor exceptions, appear correct to the nearest millimeter. Due to circumstances beyond the control of my helpers, such as the inadvisability to tamper with this model, other specifications had to calculated or extrapolated from close-up images. Such calculations, though they may not be 100% exact, do not detract from the overall image of the model or from running of the engine. All other inaccuracies are mine. There my apology rests.

Chapter I
The Rev. Robert Stirling, the Air Engine and the Stirling Engine

Facts, Assumptions and Conclusions in chronological order

1790

October 25th	Robert Stirling, son of Patrick and Agnes Stirling, was born in Cloag, Methvin, Perthshire, Scotland

1805 - 8	Studied physics and engineering at the University of Edinburgh

1809

November	Studied Divinity at the University of Glasgow

1814

November 15	Enrolled as student of Divinity, University of Edinburgh

1815

July 4th	Examined in Divinity at Dumbarton Presbytery with a view to obtaining a License to preach the Gospel

1816

September 19th	Ordained Minister and inducted to the second chapter of the Laugh Kirk, Kilmarnock. Here he had access to workshop facilities which enabled him to continue with his engineering experiments.
September 27th	Applied for and granted Letters Patent for "heat economiser" in Scotland
November 16th	Application for the same patent in England

1817

January 20th	Presented specifications to the Patent Registry
January 21st	Patent enrolled in Scotland
November 16th	Scottish patent granted

1818	***Construction of the first large scale air engine used for pumping water from a quarry based on the 1816 patent*** **(Figure 1.1)**

Figure 1.1: Stirling's original engine built around 1818, after Finkelstein

1819

July 10th Robert Stirling married Jane Rankin

Pre-1824? Robert Stirling presented the University of Edinburgh the first of his two models and some months, possibly up to 2 years later, the second model to the University of Glasgow - exact dates are not known – there appears to be no record of the presentations. (In fact it is thought that the first model was actually made not long after 1818). The presentation of the first model to the University of Edinburgh is thought to have been made while Robert was still serving in Kilmarnock.

(A label dated 1978 attached to the model bearing reference number 1975.66 stated that the model was given to the University of Edinburgh not later than 1825.)

1823 - 1825 A report in the Manchester Guardian (see 1853) indicated that a Clyde vessel, the 'Highland Lad', a 52 foot wooden vessel, single central-paddle driven, had the original steam engines replaced by a 'caloric engine' designed by Rev. Robert Stirling and constructed by Messrs Claud Girdwood of Glasgow.

(The 'Highland Lad', previously named the 'DEFIANCE', was built in 1817 and scrapped in 1827 - as described in the history of the Clyde Steamers by James Williamson, published in 1904.)

1824

February 12th Robert Stirling appointed Minister for Galston, Ayrshire. He set up his own workshop at The Old Manse where he lived. (Photos 3a, b & c)

Photo 3a (above): An early photograph of Galston Parish Church at the time of Rev. Robert Stirling (Courtesy - GALSTON Parish Church - Church of Scotland)

Photo 3b (right): A painting of Galston Parish Church at the time of Rev. Robert Stirling (Courtesy - GALSTON Parish Church - Church of Scotland)

Photo 3c: An early photograph of the Old Manse, Galston, where the Rev. Robert Stirling lived. His workshop adjoined the house on the left. He was 'often heard working late in the night by his neighbours'. (Courtesy - Henry Kerr/John Connell (GALSTON)

Figure 1.2: Stirling's 1827 double-acting air engine - the first engine to be pressurised.

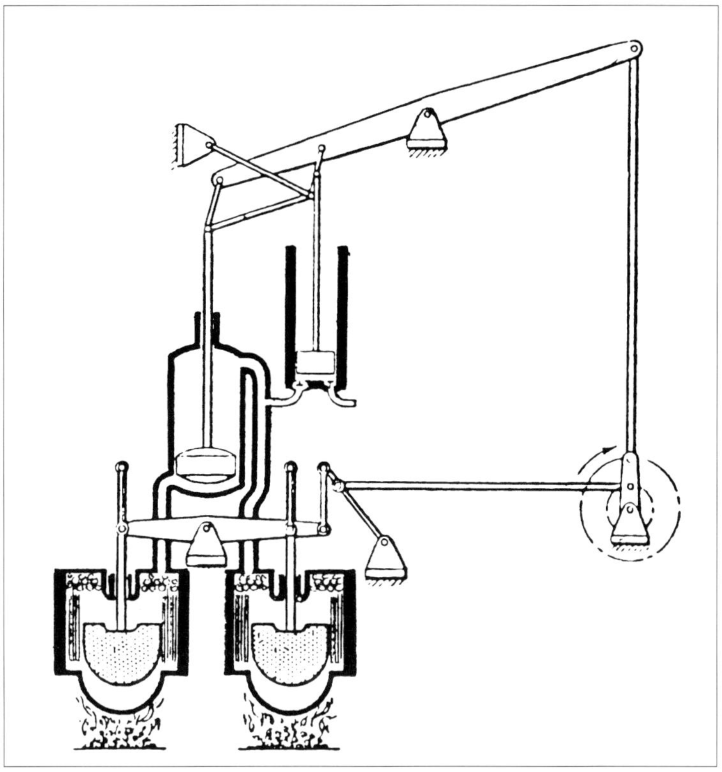

Work started on second 'air engine', based on a converted steam engine – (Figure 1.2).

1824

November 13th

A Mr. D. Screw wrote in the Mechanics Magazine referring to a previous correspondent J.Y. stated "….the machine J.Y. saw in full action is nothing more than a model of an air engine invented by the Re. Mr. Stirling". J.Y. had written comparing Stirling's air engine with Brown's gas engine. (Correspondence in Volume 3, Pages 31 – 163, October – December)

1826

Robert Stirling's Edinburgh model was used by Mr. John Leslie in a classroom demonstration. Robert Stirling presented his second model to the University of Glasgow. Eventually this model was placed in the Hunterian Museum which is part of the University and in the university precincts. (Photo 4)

Photo 4: Panel over showcase containing Robert Stirling's model donated to the University of Glasgow (Hunterian Museum, University of Glasgow)

1827

February 1st	Robert and James Stirling applied for a second patent in England (lodged on 20th July 1827)
February 13th	Application for the second patent in Scotland

{1830s – Robert and James experimented with a third engine, smaller than the previous ones, developing about 2hp with a 3 1/8" bore and a 12" stroke. However this was the most important of the three engines as it demonstrated that rejection of heat through an efficient cooling system was as important as the heating input if the engine was to provide continuous mechanical energy. Experiments with this engine led to the third patent in 1840. (Figure 1.3)}

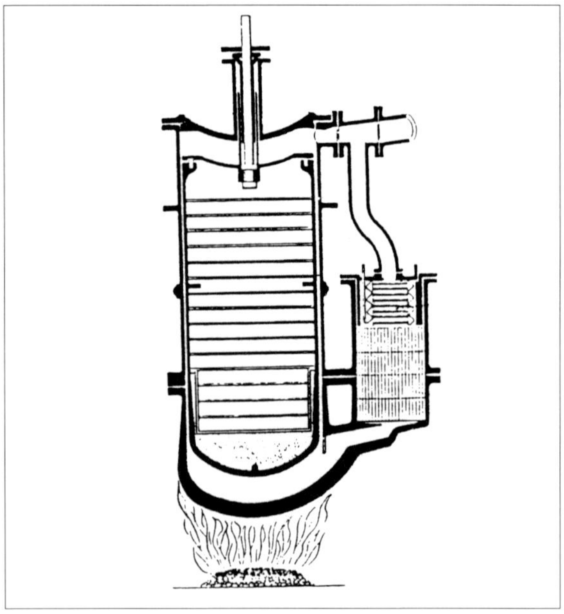

Figure 1.3: Detail from the specifications of the 1840 patent application

1840

January 11th	Robert Stirling awarded a 'Doctor of Divinity'
October 1st	Robert and James Stirling applied of a third patent in England (enrolled on March 26th, 1841)
December 17th	Robert and James Stirling applied for the third patent in Scotland (enrolled on March 17th, 1841)

Work started on a fourth engine, a 45hp double-acting displacer engine for a Dundee Foundry Company – installed in March1843 it supplied power to all the machinery of the foundry and worked continuously for nearly four years when it failed due to the poor metal, available at the time, that could not withstand constant heat at high temperature – (Figure 1.4).

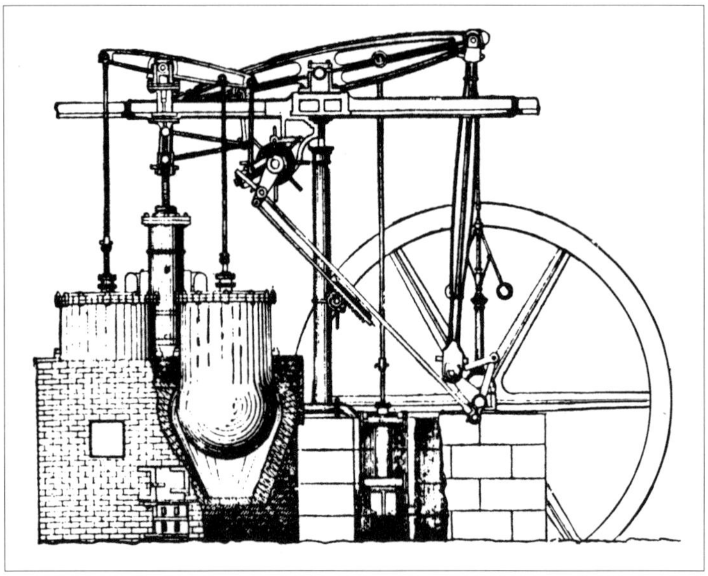

Figure 1.4: Stirling brothers' 1840 patent design on which the Dundee factory Engine was built.

1847

March 1st Prof. William Thomson wrote in a letter to J.D. Forbes of the University of St Andrews that he had discovered Stirling's 'air engine' in storage, had it taken to pieces and hoped to get it running. (Figure 1.5)

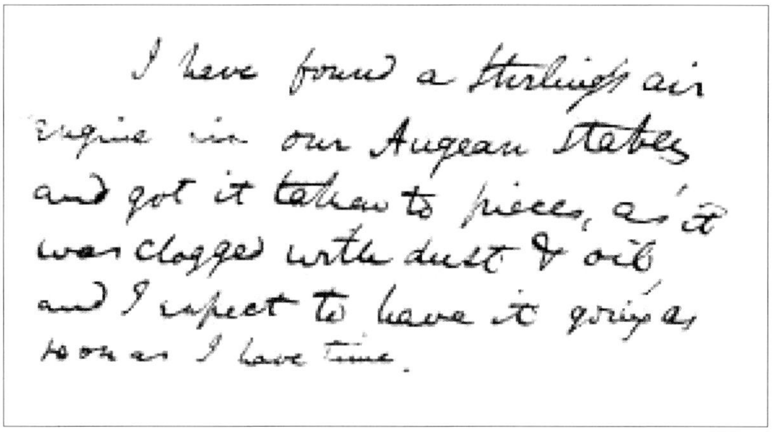
I have found a Stirling's air
engine in our Augean stables
and got it taken to pieces, as it
was clogged with dust & oil
and I expect to have it going as
soon as I have time.

***Figure 1.5:** Prof. William Thomson's description of how the Glasgow model was found in 1847. Prof. Thomson later became Lord Kelvin.*

April 21st During a meeting of the Philosophical Society of Glasgow Prof. Thomson gave an account of 'Stirling's Air Engine' and exhibited a working model. The proceedings of the meeting go on to say that… "
Attention was called to the circumstances that, in accordance with Carnot's theory...the mechanical effect to be obtained by an Air Engine, from the transmission of a given quantity of heat depends on the difference between the temperatures of the air in the cold space above and the heated space below the plunger; as this difference is considerably greater than that exists between the boiler and the condenser in the best condensing Steam Engines, it appears that, if the practical difficulty in the construction of an efficient Air Engine can ever be removed to nearly the same extent as already has been done in the case of the Steam Engine, a much greater amount of mechanical effect would be obtained by the consumption of a given quantity of fuel."

1853

February 7th An article in the Manchester Guardian commenting on the Ericsson Caloric ship stated that:
"...About 30 years ago, a vessel named the Highland Lad, that many of our readers will remember as a passage-boat on the Clyde, was fitted by Messrs Claud Girdwood & Co. of Glasgow, with a caloric engine, the invention, we believe, of the Rev. Mr. Stirling, now Minister of a parish - Galston, we think - near Kilmarnock. This engine propelled the vessel at the moderate rate that gave satisfaction in those days being estimated at 20 horses power..." Courtesy: Andrew Hall and Bob Sier

---

1876/77 Both Stirling's models exhibited in a lecture/demonstration in London. The Glasgow engine was described as in 'working order'.

1878

June 6th Robert Stirling died in Galston, Scotland, at the age of 88 years (Photos 5 & 6).

1917

September 6th Original copy of the Rev. Stirling's English patent specifications presented to the London Patent Office library by Rev. Stirling's daughter Agnes Hunter, after these were found among some old documents.

In the intervening years between 1840 and 1880s there were many developers who attempted to obtain greater efficiency from the caloric engine and a good number of patents were awarded. However by the early 1900s very few large hot air/caloric engines were being manufactured as the steam engine gathered more support with various safety measures and better materials. Some of the caloric engines that remained popular for a few years were those manufactured by A.E.H Robinson, Rider/Ericsson, Heinrici, W.H Bailey, Lehmann, Van Rennes, Gardner & Sons. With the advent of the internal combustion engine and of electricity the fate of the 'air engine' as a prime mover seemed to be sealed for good – only toys were being manufactured by 1930.

Photo 5: Rev. Robert Stirling's headstone in Galston Public Cemetery - it includes his wife, Jane, (nee Rankin) and his elder daughter, Jane.

Photo 6: Inscription on the pulpit of Galston Parish Church erected by the Stirling family in 1913.

Chapter 2
The Stirling (or Air) Engine

Rev. Robert Stirling's models donated to the Universities of Edinburgh and Glasgow have a basic, simple design, easy to run and capable of giving a creditable performance even after many years of storage or on display. Yet this simple design embodies three fundamental requisites which all modern Stirling engines have:

- a closed-cycle operation;
- an internal regenerative structure
- heated and cooling areas separation.

At least four engines are known to have been built by Robert assisted by his brother James. The second one, in 1827, was the first attempt at pressurising an engine. To gauge what foresight Robert and his brother James had, the next serious attempt at pressurising engines was made by Philips of Eindhoven in 1943.

An over-view of how the 'Stirling' or 'air' engine works: (the following is a simplified version of the cycle that occurs during the running of the engine).

The main components of the models described in this book are:

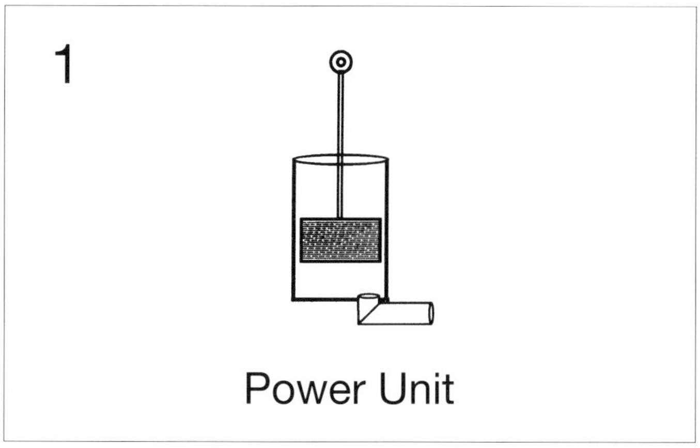

1 The WORKING CYLINDER or 'power cylinder' and piston that exerts force and momentum to the flywheel causing it to turn;

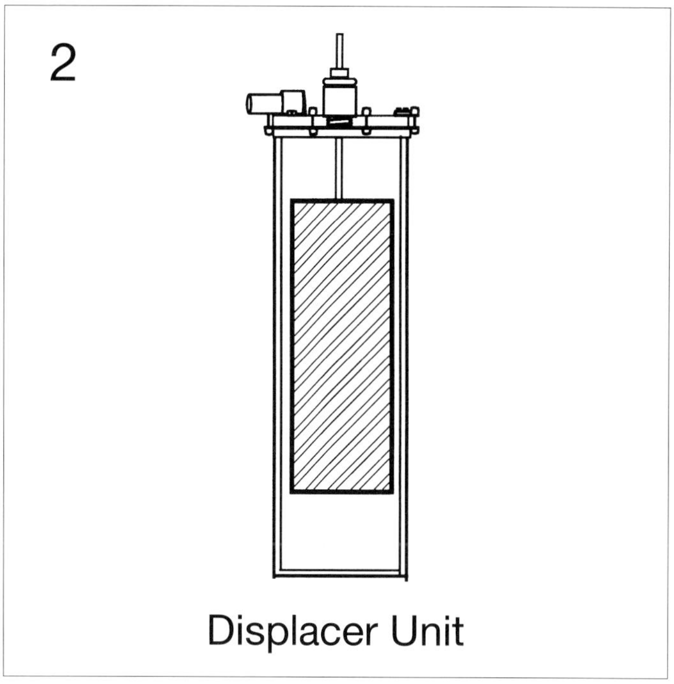

2 The DISPLACER CYLINDER and DISPLACER (also known as the 'plunger'). In this cylinder a pocket of air (gas) is heated and cooled alternately when the plunger or displacer moves it from one end to the other;

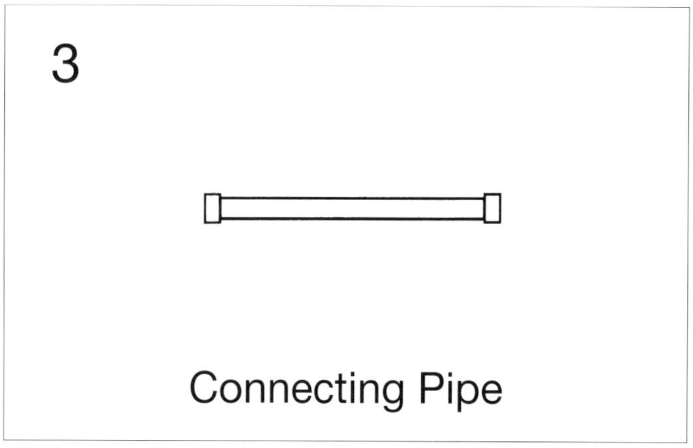

3 The CONNECTING PIPE between the power cylinder and the displacer cylinder which allows the transfer of air pressure;

Flywheel

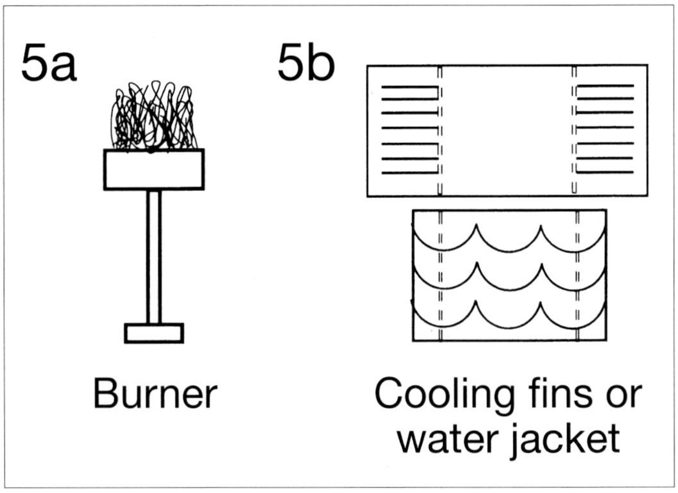

Adjuncts

4 The FLYWHEEL – which provides momentum to keep the engine running; this is connected to the power piston and the displacer;

5 HEATING AND COOLING (fins or water jacket) devices that are essential for the engine to run.

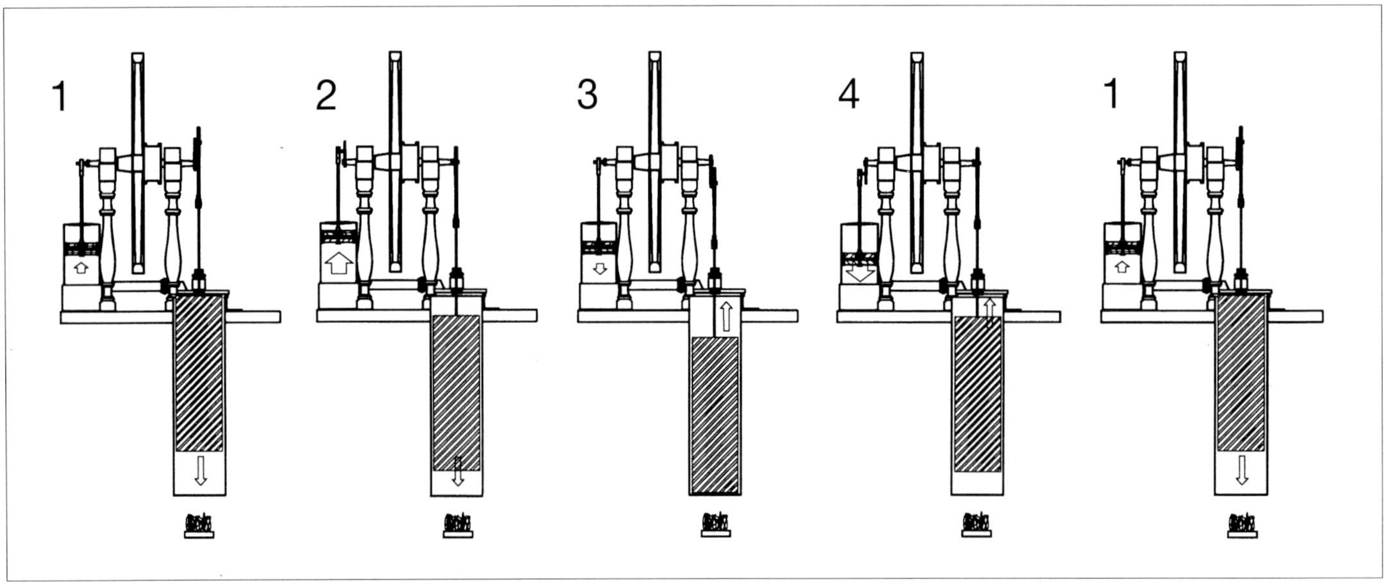

Sequence of movement of the displacer (plunger) in relation to the power (working) piston

In the simplified set of diagrams shown above:

in **1** one finds a pocket of air or gas in the lower part of the displacer cylinder. When heat is applied, the volume of air expands and pressure rises. The elevated pressure moves past the displacer into the power cylinder causing the power piston to move upwards and the flywheel to turn;

in **2** one finds that the power piston has moved to the maximum of its stroke, the flywheel has turned quarter of a revolution, the displacer has moved downwards pushing a volume of air in the upper (colder) area of the displacer cylinder where it starts to cool;

in **3** one finds that the flywheel momentum (aided by a partial vacuum caused by air being cooled) has caused all the air in the displacer cylinder to move into the upper (cold) area where it cools rapidly, while the piston is on its way down to the original position;

in **4** one finds that the flywheel momentum and the power piston downward movement have caused the air in the upper part of the displacer cylinder to move down to the heated area where it starts to heat, expand and cause increased pressure until.......

the cycle is completed as in diagram 1 where ALL the air is now being heated resulting once more in an increase of air pressure.

This explains the more important and basic theory that in the 'Air' engine, heat is converted into mechanical energy.

To enable the cycle to operate well, two other important factors are required:

i the power piston and displacer are connected to the flywheel at 90° or a quarter of a turn, to each other, one in advance to the other;

ii while the power piston is airtight (gas tight) but with minimum friction, the displacer (or 'plunger') is loose fitting, that is having a narrow (annular) gap between the displacer and the inner wall of the cylinder, which allows air to move in either direction, from one end to the other. The gap is generally very narrow, preferably not more than 1mm maximum in small engines.

What is also very interesting in the process is that the displacer itself is generally hot at the bottom and cold at the top, with a heat gradient along its length. Thus when air is moving down to the heated end, it collects heat in a gradual manner and as a result it becomes hot more quickly when it reaches the bottom.

Well over a century after Stirling's first patent Philips of Eindhoven, Netherlands, undertook serious studies and development of the closed-cycle air engine which then became synonymous with the Rev. Robert Stirling and the coined description for this type of engine became the 'Stirling cycle engine' or more briefly, the 'Stirling engine'

CLOSED-CYCLE vs OPEN-CYCLE OPERATION

Two types of air/hot air/caloric engines were developed during the early years of the nineteenth century, the 'closed-cycle' and the 'open-cycle'. Briefly explained, in a 'closed-cycle' operating air engine the volume of gas (air) in the engine compartments is trapped and not allowed to escape. Moreover this enclosed mass of air is made to work by expansion and contraction with every cycle, that is a complete revolution of the flywheel, thus re-circulating continuously.

There are various types of 'closed-cycle' air engines, but basically the two more common ones are either with two cylinders with a connecting tube or pipe or with one cylinder with two pistons inside. The models that Robert Stirling presented to the Universities are of the first type, however the engine which he presented in the specifications attached to the 1816 patent was of the second type, that is one cylinder with a power piston and a displacer. This latter type, which is sometimes called a concentric engine, is the more efficient since there is very little dead space between the two pistons and the volume of air is made to be more productive. The very fact that all the powerful air engines built in recent years are closed-cycle concentric design proves Robert Stirling's foresight, a foresight that was not acknowledged for decades simply because the basic principles of thermodynamics were not known and when these were published they were not understood by many developers.

'Open-cycle' air engines operate with a fresh charge or induction of air by means of valves. These engines were more noisy, less powerful and prone to failure due to faulty, leaking or burnt valves and seating. Such engines were at first closely associated with John Ericsson and although many other air engine developers followed this system none really managed to achieve long lasting production. In due course Ericsson also soon joined the band of the 'closed cycle' developers

The Regenerator

The 1816 patent is based mostly on the principle of 'economising of heat'. The premise of the Robert Stirling was that heat could be extracted, stored, released and re-used during each cycle of operation, thus after the initial amount of energy is used to bring heat to the required level for running the engine, further application of heat can be reduced as the 'returning' air is already pre-heated and requires only 'topping up'.

It should be explained that air is moved from the hot and cold ends along a narrow radial space between the outer cylinder and the moving inner cylinder known in Stirling's days as 'transferer' or 'plunger' and now termed as 'displacer'. This narrow radial space, according to Stirling, could be closely filled with heat absorbing material such as metal strips, glass, ceramic, etc. In fact the 1816 drawing with the specifications shows a mass of coiled wire along the gap between the two cylinders.

The term 'regenerator' came into use much later.

Stirling engine technology

As a result of Stirling's foresight and brilliant mind ably demonstrated in the 1816, 1827 and 1840 patents granted to Robert and his brother James, a technology evolved which laid the basis for the study of thermodynamics as we know it now. It was as a result of studying Stirling's models that Prof. William Thomson, later Lord Kelvin, established the Absolute Scale of Temperature, coined the word energy and invented the new science of Thermodynamics.

The Stirling engine technology can be summarised with the following 'principal' characteristics:

- **The Stirling engine is an externally heated engine (external combustion) in which a volume of gas (air, helium, hydrogen etc.) is enclosed and re-circulated during each complete revolution of the flywheel (i.e. a complete cycle).**

- **A Stirling engine works by compressing and expanding an enclosed volume of gas at different temperature levels.**

- **Any source of heating (multi-fuel) can be used to run a Stirling engine.**

Photo 7: Philips multi-fuel demonstration Stirling engine - capable of working equally well on alcohol, olive oil, salad oil, lubricating oil, two types of fuel oil, diesel oil, kerosene and petrol.

The versatility of the Stirling engine which was ably shown by a demonstration Stirling engine which was made to run of several different fuels, from salad oil to olive oil, alcohol, kerosene, petrol, diesel oil, fuel oil and lubricating oil (Photo 7).

- **A Stirling engine can achieve high efficiency if a good and effective regenerator is used.**
- **The Stirling engine is versatile in design and in the types of drive mechanisms, an advantage which very few prime movers have.**
- **The Stirling engine requires minimum lubrication, yet wear and tear is negligible.**
- **The Stirling engine is virtually silent as it has no internal combustion or explosion of fuel, such as happens with petrol, diesel, gas engines.**
- **The Stirling engine is one of the most environmentally friendly engines yet designed with no exhaust and very little emission.**

Other advantages include:

- **The Stirling engine is more economical to operate.**
- **There are fewer moving parts than in most other engines.**
- **There is less stress on the moving parts giving longer life to the engine.**

Dr. Ir. R.J. Meijer, one of Philips most well-known developers in the Philips Research Laboratories, in his paper on 'The Philips Stirling Engine as a Propulsion Engine' (1969) wrote: *"there are ….. some twenty independent variables which together affect the power and efficiency of the Stirling Engine".*

David Urwick, a well known engineer and serious student of the Stirling engine, (my mentor in Malta, 1978 - 80), wrote as follows in ***'Model Engineer – 1975'***: *"The attraction of the hot air engine is that the variables are so numerous that the scope for invention and mechanical experiments has by no means been exhausted".*

A paper presented at one of the I.S.E.C. meetings by Naoutsugu Isshiki, Prof. Emeritus of Tokyo Institute of Technology with the subject 'R&D at the Ship Research Institute' states: *'In 1975 several senior academics, including the author, representing different Japanese Universities visited Europe to familiarize themselves with the research into Stirling Engines taking place in Sweden, Germany, Netherlands, etc. The members of the group were very impressed by the scale of ongoing activities in this area and the outcome from the trip was that the Japanese Ministry of transportation initiated a large scale project on Stirling Engine Technology.'* Japanese academics and engineers are well known for their deep research and absolute conviction in any project that could be developed further. The enormous research being undertaken by several universities and technical colleges augurs well for new technological advances for the Stirling engine.

Now, thirty + years later, never have words proved so true. The development being carried out by major firms and studies being conducted in numerous universities world wide are aimed at producing powerful, economical and silent Stirling engines for use in many walks of life.

CHAPTER 3
THE REV. STIRLING'S MODELS OF AIR ENGINES

(For ease of reference in the following chapters and to indicate more quickly one or other of the two models, the author henceforth will call these models as the EDINBURGH model and the GLASGOW model!)

The models built by Rev. Robert Stirling possibly with some help by his brother James have a basic, simple design. They are elegant in a way that captures attention. Robert Stirling could have designed the engines in a half-a-dozen different configurations, the most common being lengthwise with the cylinders in a horizontal position – but no other position would have been as attractive as the one he chose. The use of rich reddish-brown wood and the delicately turned decorative columns accentuate the elegance of these models.

The position of the flywheel, with its off-centre and lofty position, balances and compensates for the elongated cylinder beneath the engine platform, while the shorter cylinder to the left side of the flywheel is not incongruous. The focus of attention during the running of the engine is definitely on the flywheel. The eight turned columns – four supporting the flywheel and four supporting the engine platform as well as other cross-bars at the bottom complete a graceful and captivating picture.

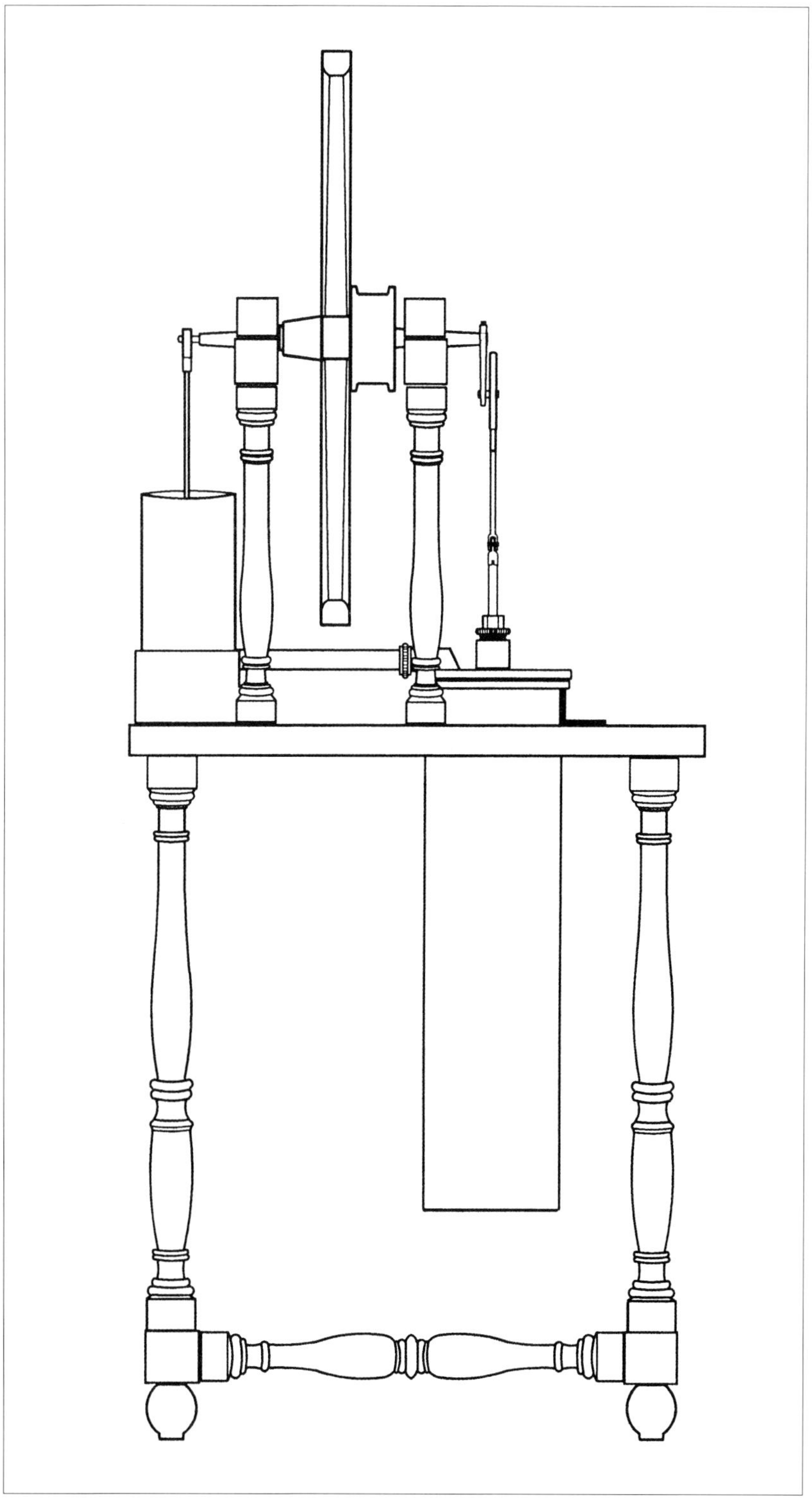

Fig. 3.1: Robert Stirling's model of the 'Air Engine' held in the Royal Museum of Scotland, EDINBURGH

The Edinburgh model (Figure 3.1)

The engine stands 29" (73.7cm) high, with a 12" (305mm) flywheel supported on two lintels by four 7" (178mm) turned wood columns. The wood engine platform is 12½" (317mm) long, 6" (152mm) wide and 11⁄16" (17.5mm) thick. The platform is in turn supported by four turned columns of elegant design resting on two lateral stands (cross-bars) on four rounded ball-feet. Two horizontal turned wood balusters hold the bottom structure firmly together. This structure was additionally strengthened by two wire ropes or strands.

The long cylinder below the base is of course what we now term as the displacer cylinder with an internal displacer (originally called 'transferer' or 'plunger') connected to the crankshaft by means of a wrist joint and connecting rod/crank-pin.

The small cylinder on the left hand side of the flywheel is the power (working) cylinder that transmits force to the flywheel. This cylinder is connected to the displacer cylinder mentioned above by a pressure transfer pipe which enters the base of the

Photo 8: The author's half-scale replicas of Stirling's models.

power cylinder from the sealing flange of the displacer.

The Glasgow model (Figure 3.2)

Rev. Stirling's Glasgow model is, in the opinion of the author, a later version, maybe by a few years. The visible differences as explained above can be condensed as follows:

- **Different elongated shapes of the top and bottom columns, still elegant, but requiring less workmanship;**
- **Different style of horizontal bearing blocks securing the flywheel;**
- **The square ends of the horizontal cross-bars supporting the bottom columns;**

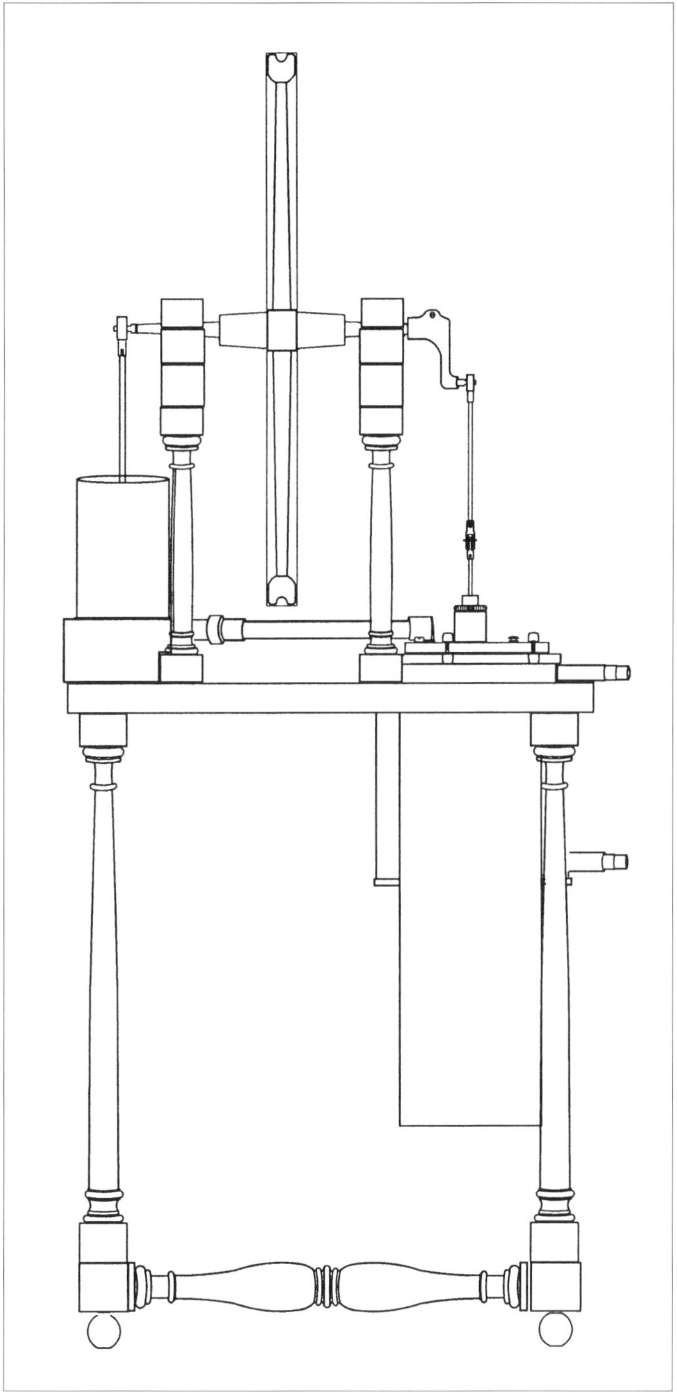

Fig. 3.2: Robert Stirling's model of the 'Air Engine' held in the Hunterian Museum, University of Glasgow, SCOTLAND

- **Absence of a pulley on the crankshaft;**
- **The shape of the flywheel which has a grooved rim to act as a large pulley;**
- **The displacer cylinder is probably cast iron, but could be of sheet iron – unlike the Edinburgh model which was made from brass.**
- **The wrist joint connection between the displacer crank and the displacer con-rod and**
- **The important addition of a water cooling jacket.**

All other measurements are practically the same as the Edinburgh model with minor differences, one of which is the diameter of the flywheel which is 12⁹⁄₃₂" (31.2cm). The Glasgow model also stands 29" (73.7cm) high.

CONJECTURES on the important differences between the Edinburgh and Glasgow models

Of the visible differences mentioned above most are cosmetic but the last three have a particular importance. The difference in woodwork and flywheel/pulley layout could have been as an expression of avoiding a carbon copy arrangement. However the addition of a water cooling jacket has a significant meaning and this has to be seen in relation to the work and exposition of theories by Professor William Thomson, later Lord Kelvin.

It could have been that Robert Stirling realized after running his first model that heating and cooling separation in the Edinburgh model was not sufficient and as a result the engine lost power after it had been running for some time. It was then, therefore, that he included the water cooling jacket for the Glasgow model. This should not be surprising because even in his application for a patent in 1816 (English Patent No. 4081 of 1816), Robert Stirling included cooling as follows '....and I apply a stream of cold air or water to the coldest part of the engine to carry off waste heat'. Further on in the application he mentions that the difference in temperature between the hot and cold ends should be '480° (F) higher than part...'. This therefore proves that even in his first experiments, Stirling realised the importance of cooling (by water). Although Stirling's patent application was basically that 'of saving fuel' his 1816 design and description contained features which were not really understood for over a century and were only incorporated fully in the Philips researches around 1936.

On the other hand when Professor Thomson found the abandoned Glasgow model 'in our Augean stables, and got it taken to pieces....' some parts of this model were replaced, indeed some were replaced more than once. It is not beyond comprehension that the addition of a water cooling jacket was undertaken while the engine was dismantled and some parts renewed. Professor Thomson emphasized on the extraction of heat by 'passing a stream of water... around the cold end'

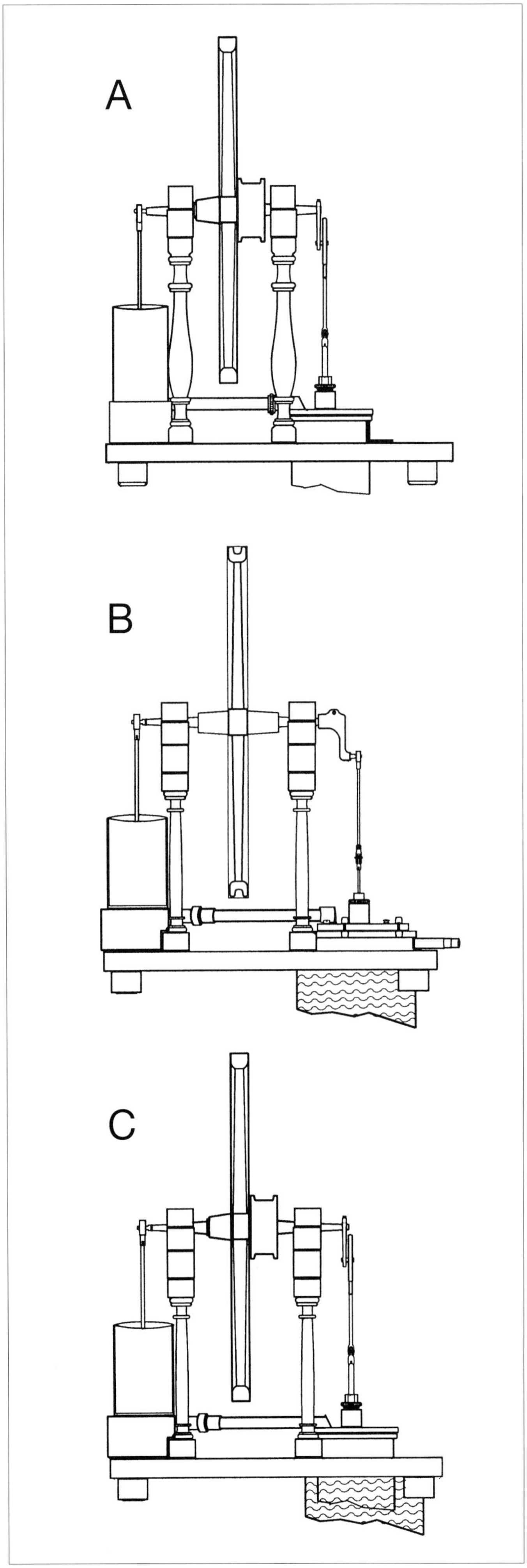

both in classroom demonstrations as well as during his 'live' demonstration during a meeting of the Philosophical Society of Glasgow on April 21st 1847. Professor Thomson went even further by explaining that if the model was reversed by the use of an external force the lower part of the cylinder, which is normally heated to run the engine, becomes cold to the extent that a sheet of ice forms, while the upper part retains ambient temperature.

A comparative study of the mechanisms of the two models 'A' and 'B' in Figure 3.3 seem to indicate that both were designed in approximately the same layout such that a super-imposition of the Edinburgh model on the Glasgow frame has an almost exact fit as seen in montage 'C'. This seems to suggest that the water cooling jacket was added on at a later stage. To add credence to this view, the pedestals of the top right hand columns and the capitals of the bottom right hand columns have had their corners cut at an angle to accommodate the water jacket. If this cooling device had been designed before construction, it can be rightly assumed that Robert Stirling would have taken steps to adjust the mechanism layout slightly different and the woodwork would be untouched.

One other casualty of the rehabilitation of the Glasgow model is the power unit. Here also the base of the power cylinder appears to have been changed and a slightly larger one installed. Whereas in the Edinburgh model the power unit (and base) lie snugly between the left hand top columns, in the Glasgow model the wooden base was made wider with the result that to fit it in place the front end was reduced in width. Even so the base (and the power unit) extend beyond the engine platform giving it a rather untidy finish.

We know from first hand information given by the Hunterian Museum that the engine platform had a slot cut to accommodate a fastening screw from the wooden base of the power cylinder to allow it to slide in order that the new connecting socket between the power cylinder and the displacer cylinder could be soldered. Whereas the slot and fastener are hidden beneath the platform, the soldered connection is clearly visible when the engine is viewed closely.

The displacer crank has a modern appearance and is totally incongruous with the historic appearance of the model, and it is a wonder how and why a better fitting could not have been machined. The power crank has also been replaced, but this at least fits slightly better.

The wrist connection between the displacer crank and the connecting rod is straight-forward, eliminating the slotted yoke of the Edinburgh model. Whether this is due to the rehabilitation of the model, or because Robert Stirling opted to give the stroke the correct length, is also subject to conjecture. The same applies to the power crank/con-rod connection which is different from the Edinburgh model in that the latter had two settings for the stroke of the power piston.

Figure 3.3: Comparison between the layout of the Edinburgh and Glasgow models

Chapter 4
Scaling Down the Stirling Models (and other problems arising)

The construction notes given in this book are for a full scale replica, and while most of the designs are shown in 50% scale, the specifications alongside are for the original size of the models. For some small components, such as cranks, yoke, clevis etc the drawings are full scale to assist in the machining process. One or more have been increased even further. Appendices 1 and 2 give the specifications for full-size replicas of Robert Stirling's models. Photographs taken by the author are of a 50% scale model, unless specified otherwise.

Stirling enthusiasts who would prefer to make a full scale replica will find it easier to do this than a half-size model since some of the small components are more manageable than when reduced to half size. Any machinist who has no problem in displaying a model almost as tall as a table, nor of carrying it around, will find it relatively easy to follow the construction notes together with the specifications in the appendices.

Readers who opt to construct half-scale models, as the author has done, have only to halve the specifications and get on generally well with the same construction notes. Scaling down brought about a few problems due to the size of some components. The delicately turned columns or pillars, reduced to half size, presented some difficulty as the author's lathe and tooling were not quite suitable for such small delicate wooden components as the top four pillars. Copying four columns compounded the problems, although in the end a jig made from bright mild steel did a fair job. Wood was another problem as although mahogany is fairly hard, turning and contouring a very narrow diameter (between ⅜" = 0.375" (9.5mm) for the top columns and 17/32" = 0.531" (13.5mm) for the bottom columns, mahogany tended to flake. In the end beech was used for the turned columns.

The power and displacer cranks also presented a problem. For example, the thickness of the original power crank tapers from 13/64−" = 0.2" (5.08mm) to 5/32+" = 0.16" (4.15mm). A 50% thickness gives a rather weak hold on the crankpin. The displacer crank is only marginally thicker while the yoke attached to the displacer rod is 3/16" = 0.187" (4.76mm) thick and has a 5/32" = 0.152" (3.88mm) slot. Reducing by half all these components may cause problems during long engine runs.

Other problems arising

Now, after almost two centuries after Robert Stirling's first patent, we find that Stirling engines are generally deemed to be efficient if the following two parameters (among the many others) are met:

- **A ratio of 1.5:1 between the swept volume of air in the displacer cylinder and the swept volume in the power cylinder;**

- **a phase angle of 90° between the stroke of the displacer and that of the power piston.**

Robert Stirling's models in the Royal Museum of Edinburgh, in the Glasgow University Hunterian Museum and the replica in the London Science Museum all have a 90° phase angle between the two cranks.

From calculations made of the models, it results that the ratio of swept volumes is much higher than norm, the Edinburgh model to the extent of 3.5:1, that is the swept volume of air in the displacer cylinder is almost four times that of the power cylinder, while the Glasgow model's ratio is slightly less, 3.4:1. Any engine constructed by the author to any of these ratios would not have worked, or if it did, it would have been erratic and very low in power.

The author's guess is that the high ratio was compensated by a large volume of dead space in the pressure pipe between the two cylinders. Materials used in Robert Stirling's model, particularly brass, leather-covered power piston and tightened cotton/leather strip in the glands could also have been a contributing factor.

It has been confirmed and documented by letters and reports that both models made by the Rev. Robert Stirling ran efficiently when many years later they were used in class room demonstrations as well as some fifty years later, in 1876, when the Glasgow engine was shown (and described) as 'running sweetly' during a lecture/demonstration in London.

Earlier in this chapter mention was made of the scaled designs: most are in the 50% scale, some (like the big wooden structures and the assembled models) had to go down to 25% scale and even smaller, while a few which were deemed important for the machinist are either in the 100% or even 200% scale). Those that differ from the 50% scale carry a qualifying note.

Chapter 5
Where to Start and Materials to Use

As described before there is a certain elegance about the Rev. Stirling's models – the visual appearance is enhanced by the extensive use of dark reddish brown wood particularly with the finely turned top and bottom columns. The position of the flywheel on the four top columns is eye-catching.

It has been difficult to reach any firm conclusion on the type of wood used in Stirling's models, even experts differ since tests cannot be carried out. Beech and oak were probably used, possibly also mahogany. That more than one variety was used on each model is probable for what was suitable for the columns may not have been so for the engine platform and the crossbars. The shrinkage and warping of the Glasgow model, due no doubt to the storage conditions in the early years, seem to indicate that some type of wood was more susceptible than other. Different hues attest to the fact that at least two types of wood were used on the Edinburgh model which is in a better condition.

In the following construction chapters reference is made to mahogany and beech, but it is up the the modeller to identify which type/s of wood to use - conditions country-wise and climate-wise have an influence on the durability of the completed models.

The layout of these engines caused some moments of reflection on how to proceed with the construction sequence; in the end it was decided to start with the woodwork so that when the engine components were machined, these could be rigidly mounted for testing and preliminary running.

Readers may opt for their own sequence in constructing the various parts, however the processes explained in the following chapters can be taken out of sequence and adopted to any system as the reader desires.

The following sequence was followed in making the ½ scale replica:

Chapter 6. WOODWORK - Edinburgh Model:

1 Engine platform,
2 Four top turned columns or pillars,
3 Lintels and flywheel securing brackets,
4 SUB-ASSEMBLY – Stage 1,
5 Four bottom turned columns,
6 Two balusters (horizontal turned bars),
7 Two base cross-bars and ball-feet,
8 SUB-ASSEMBLY – Stage 2,
9 Power cylinder platform,
10 Pulley,
11 Fillet.

Chapter 7. METAL COMPONENTS - Edinburgh Model:

12 Machining of flywheel,
13 Machining of crankshaft and bearing shells,
14 SUB-ASSEMBLY – Stage 3,
15 Machining of power cylinder, pressure transfer pipe and fittings, power piston, clevis, con-rod and fittings, power crank and pin,
16 SUB-ASSEMBLY – Stage 4.
17 Displacer cylinder, cylinder head, displacer rod bush/gland, pressure pipe outlet,
18 Displacer, displacer rod and clevis,
19 Displacer crank, crank pin, yoke and con-rod,
20 FULL SCALE ASSEMBLY and TESTING,
21 Burners.

Chapter 8. ROBERT STIRLING'S MODEL OF THE AIR ENGINE AT THE HUNTERIAN MUSEUM (UNIVERSITY OF GLASGOW) – INTRODUCTION

Chapter 9. WOODWORK - Glasgow Model:

1 Engine platform,
2 Four top turned columns,
3 Lintels and flywheel securing brackets,
4 SUB-ASSEMBLY – Stage 1,
5 Four lower turned columns,
6 Two balusters (horizontal turned bars),
7 Two base cross-bars and ball feet,
8 SUB-ASSEMBLY – Stage 2,
9 Power cylinder platform,
10 Fillets.

Chapter 10. METAL COMPONENTS - Glasgow Model:

11 Machining of flywheel,
12 Machining of crankshaft and bearing shells,
13 SUB-ASSEMBLY – Stage 3,
14 Machining of power cylinder, pressure transfer pipe, power piston, clevis, con-rod and fittings, power crank and pin,
15 SUB-ASSEMBLY – Stage 4,
16 Water cooling jacket, displacer cylinder, cylinder head, displacer rod bush/gland, pressure pipe outlet,
17 Displacer, displacer rod and clevis,
18 Displacer crank, crank pin and con-rod,
19 FULL SCALE ASSEMBLY and TESTING – Stage 5,
20 Burners,
21 Cooling arrangement,
22 Running the 'Rev. Robert Stirling's Models'.

CHAPTER 6
WOODWORK - EDINBURGH MODEL

CONSTRUCTION NOTE (APPLICABLE TO EDINBURGH AND GLASGOW MODELS):

Measurements are given in fraction inch, decimal inch and metric. The metric and decimal inch are more accurate than the fraction inch. Readers will note that some fraction inch measurements have an additional (+) or (–) – the difference is slight, sometimes by less than 1/128" which does not effect either the layout, design or function of the models.

READERS ARE REMINDED THAT THE AUTHOR'S MODELS WERE CONSTRUCTED AS ½ SCALE REPLICAS – THE CONSTRUCTION NOTES GIVEN BELOW REFER TO HOW FULL SCALE MODELS SHOULD BE MADE – in the present tense!

Except where specified construction drawings are in 50% scale approximately, while the dimensions are given for the actual size of the component. In order to help machinists some small components are drawn as 'approx. size' and quoted as such.

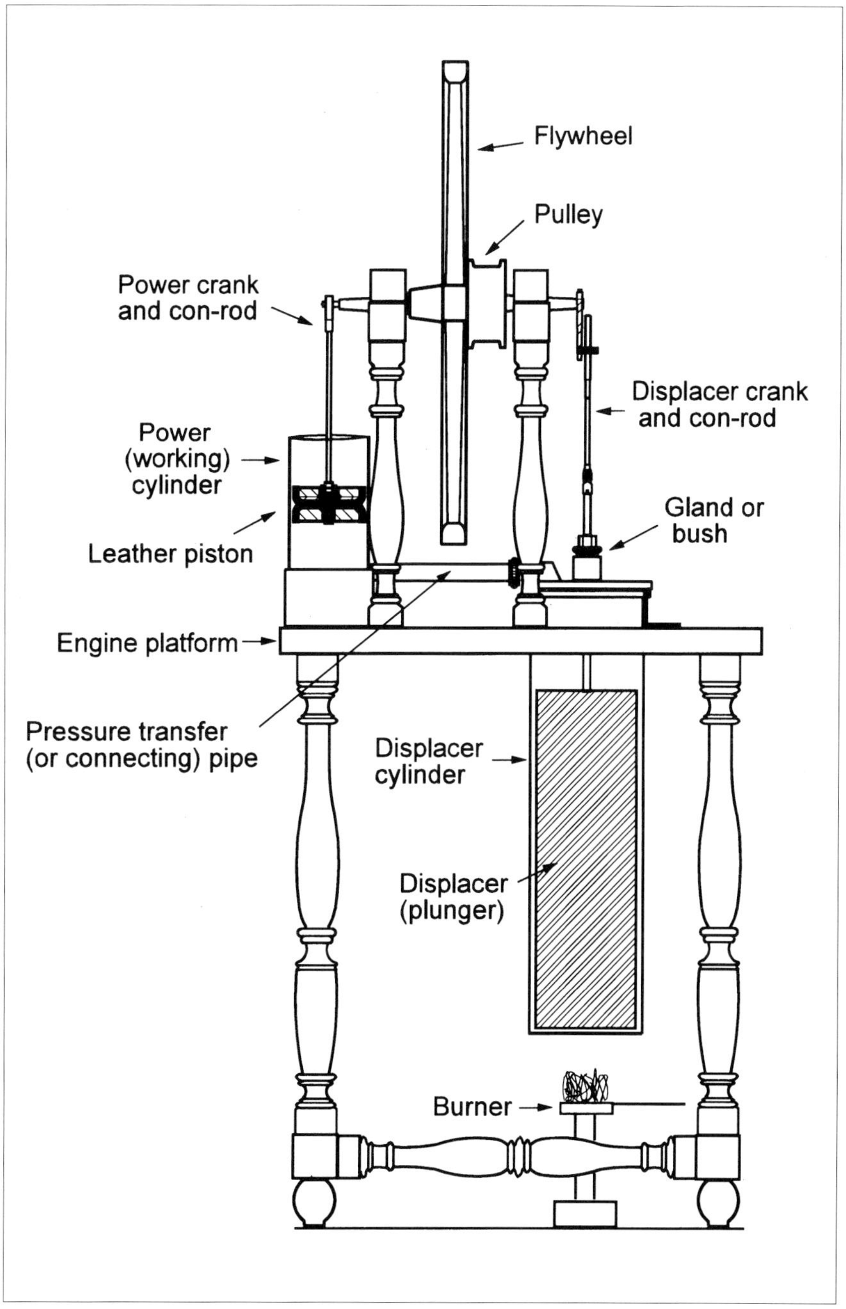

Profile of Robert Stirling's model of the 'Air Engine' held in the Royal Museum of Scotland, Edinburgh

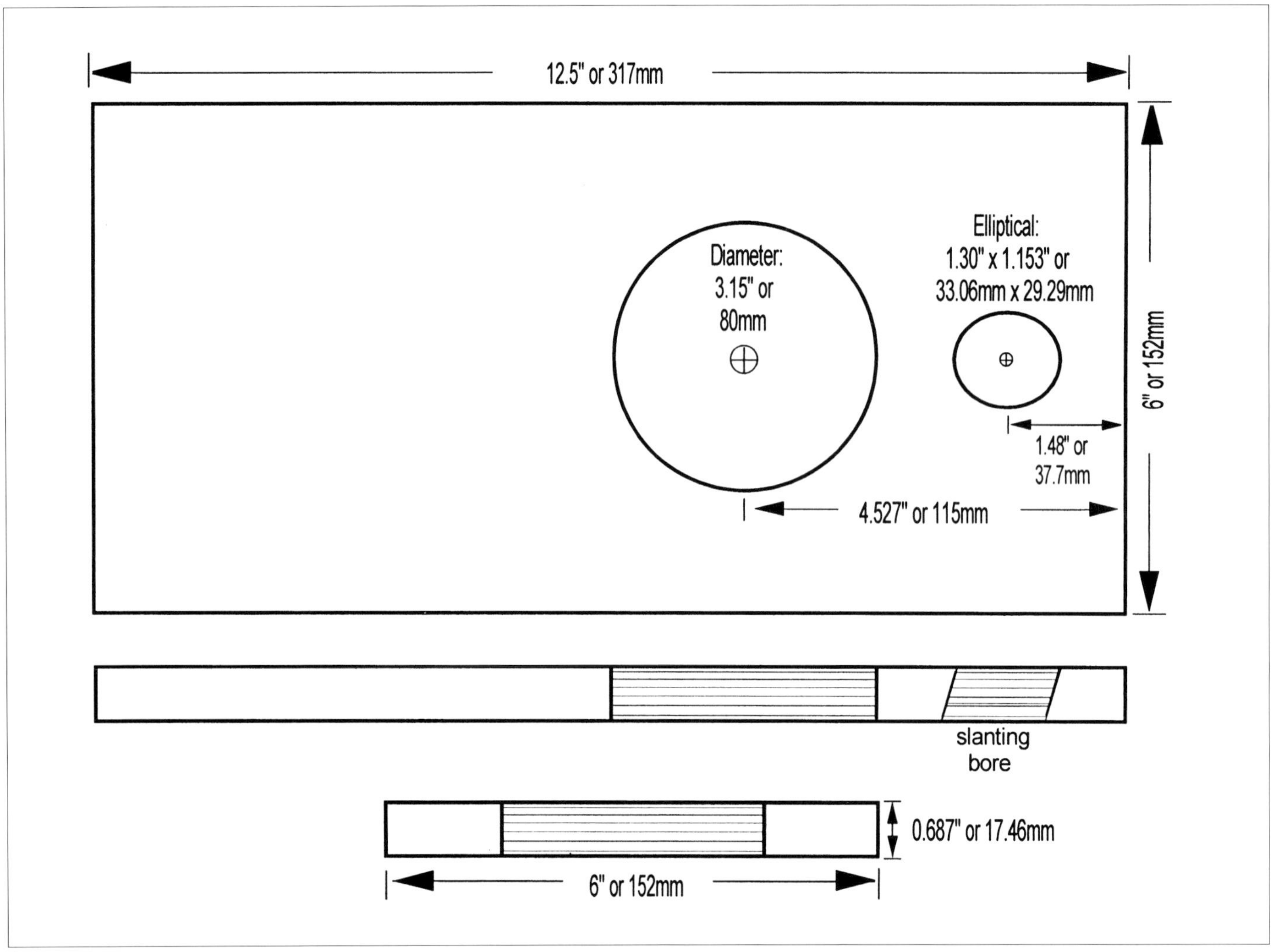

Fig. 6.1: Top view, front and side elevations of engine platform with cylinder and elliptical bores

Construction Sequence 1 – 8

1 ENGINE PLATFORM

NOTE: The author suggests that the engine platform, the position of the top columns, the bore for the displacer cylinder, the elliptical bore, the power cylinder base and the power cylinder, together with an outline of the crankshaft are drawn of a sheet of paper or cardboard before work is commenced on the engine platform.

The platform is cut from mahogany 11⁄16" = 0.687" (17.46mm) thick, 12½" = 12.5" or 317mm long and 6" (152mm) wide; two holes are bored through, one to take the displacer cylinder and the second, an elliptical one, (for which no valid reason was discovered, unless it was for a chimney flue or an outlet for a pipe to relieve heat and smoke from a burner). In fact this hole, apart from being elliptical, is also bored at an angle facing away from the engine mechanism and also slightly scorched (Figure 6.1).

The bore for the displacer cylinder is 3 5⁄32–" = 3.15" (80mm), its centre being 4 17⁄32–" = 4.527" (115mm) from the right hand edge.

The elliptical hole is 1 19⁄64" = 1.30" (33.06mm) long, 1 5⁄32" = 1.153" (29.29mm) at its widest point, with the long side in the left to right direction, its centre being 1 31⁄64" = 1.48" (37.7mm) from the right hand edge.

The platform top side is drilled in four places to take the top turned columns as shown in Figure 6.2a. - drill size 3⁄16" = 0.1875" or 5mm to a depth of 3⁄16" to take similar size dowels which also hold the top columns.

The underside of the platform is also drilled 3⁄16" or 5mm at the four corners to a depth of 3⁄16" as shown in Figure 6.2b to take lower columns. All holes are first marked and checked two ways from each adjacent edge.

The platform is sanded, given a coating of mahogany stain, and then clear semi-matt varnish. (All other wood pieces were treated in this manner to give a uniform colour.)

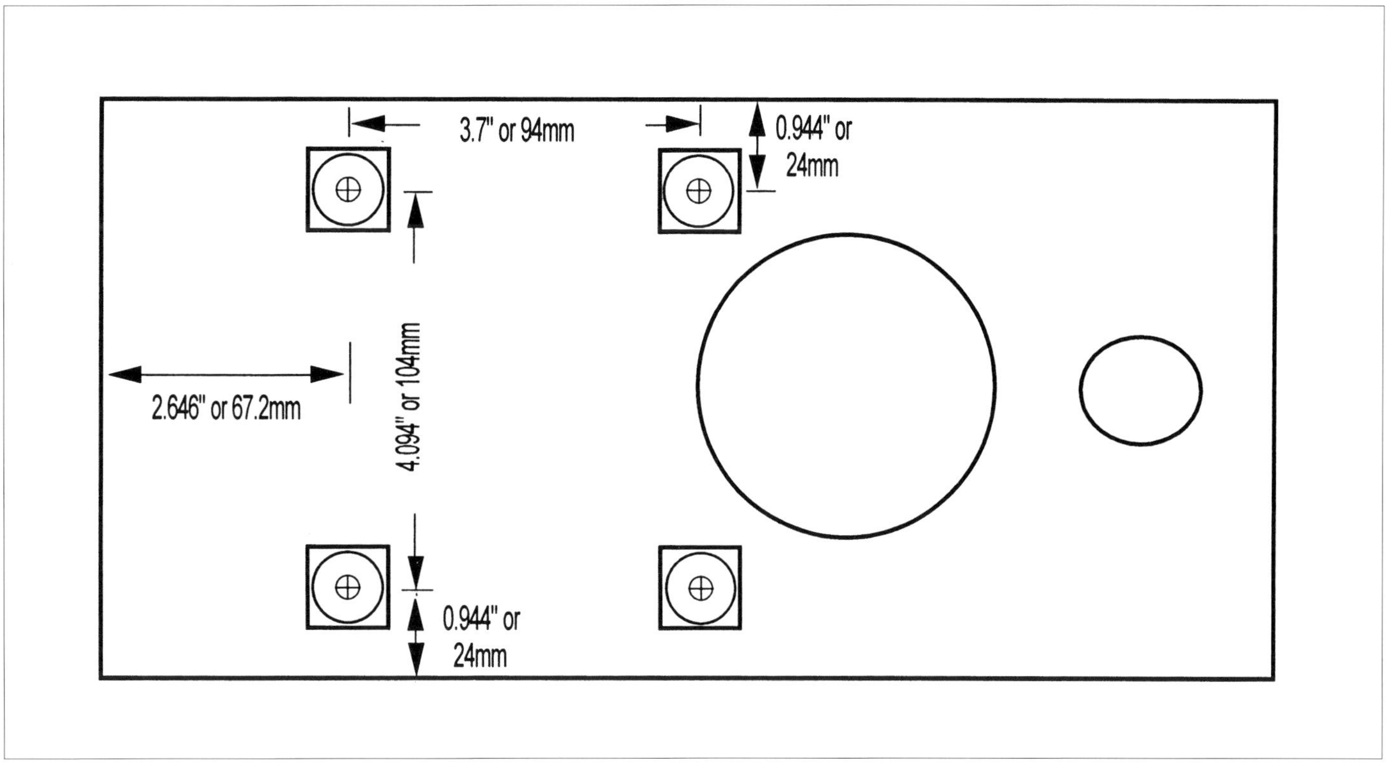

Fig. 6.2a: Position of upper turned columns on engine platform

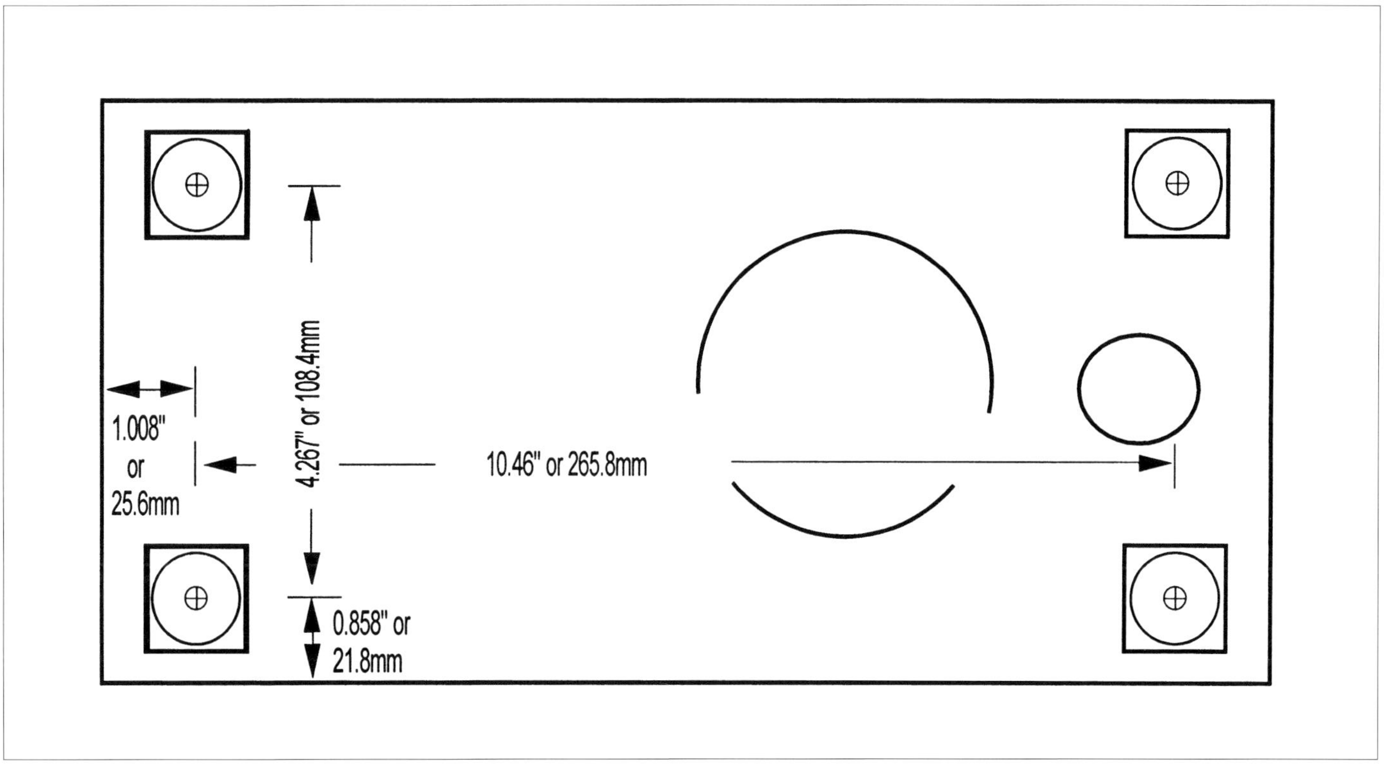

Fig. 6.2b: Position of bottom turned columns on the reverse side of engine platform

2 TOP TURNED COLUMNS

The top four turned columns are 7" or 178mm long with square capitals and pedestals 27/32" = 0.848" (21.55mm) and 5/8" = 0.625" (15.875mm) high. The wood is marked for the capitals and pedestals -16mm - from each end leaving 5¾" = 5.748" (146mm) to be shaped as in Figure 6.3, (Photo 9).

The columns are centre-drilled at the top and bottom to take 3/16"or 5mm dowels to a depth of 3/16" for fitting into the lintels and the engine platform.

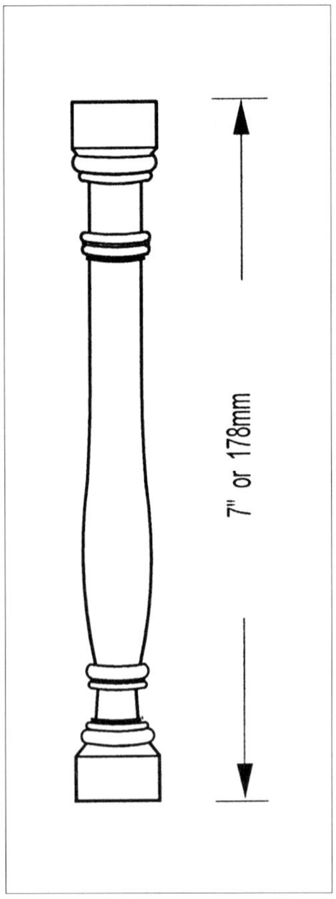

Fig. 6.3:
Top turned column (x 4)

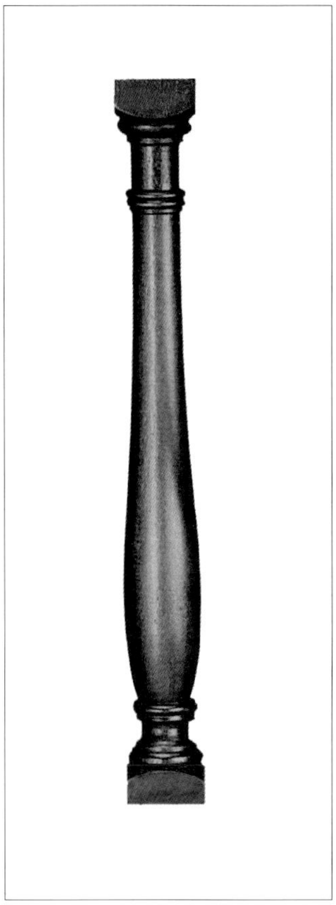

Photo 9:
Top turned column (x 4)

3 FLYWHEEL SUPPORTING BRACKETS

NOTE: ***In Robert Stirling's model the flywheel crankshaft rests between two sets of wooden brackets, each set consisting of a lintel which sits on two top turned columns, and a securing bracket which holds the crankshaft rigidly in place. The lintel and the columns were apparently glued together, however the upper securing bracket was screwed to the lintel with a set of bearing shells in between.***

The lintels are cut and finished from hardwood 5 35⁄64" = 5.546" (140.88mm) long, 31⁄32" = 0.976" (24.8mm) square. The upper securing brackets are 2 3⁄64–" = 2.04" (51.8mm) long, 29⁄32" = 0.905" (23mm) wide 51⁄64+" = 0.803" (20.4mm) high, (Figure 6.4).

Lintels are drilled from beneath to take the top turned columns with a distance of 4 3⁄32" = 4.094" (104mm) between centres as shown in Figures 6.4 and 6.5.

Each pair of securing brackets and lintels are bolted together, drilled in the centre 7⁄32+" = 0.228" (5.8mm), widened by taper reamer to 15⁄64" = 0.234" (5.95mm) on the inside to take the tapered crankshaft and bearing shells. (Bearing shells can be made from lead, copper or brass - described in Section 12. Both lintels and upper brackets are later filed down in the centre to take these shells, however this is done only after the bearing shells are pressed and their place on the lintels/ brackets marked and then filed such that the wings of the shells are flush with the wood).

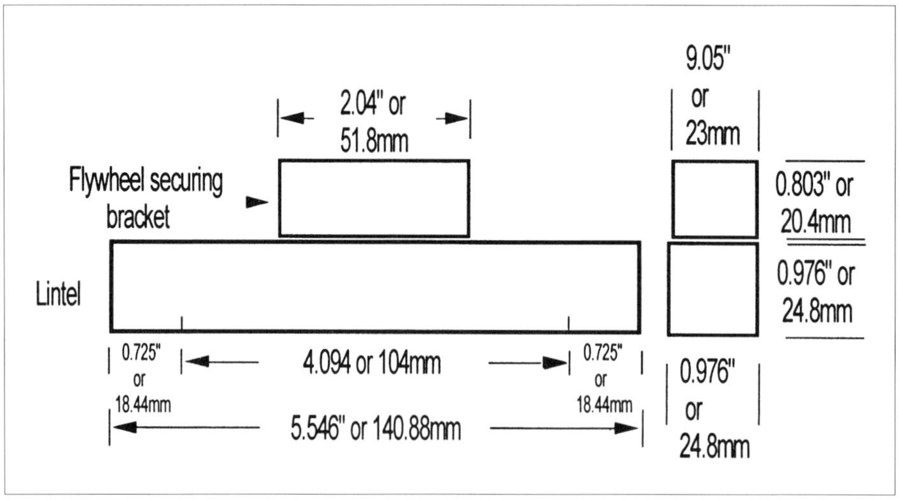

Fig. 6.4: Lintel and flywheel securing bracket (x 2)

4 SUB-ASSEMBLY - STAGE 1

As a first step a trial assembly is made of the top four columns - first into the platform and then into the lintels with a suitable rod acting as a temporary crankshaft secured by the top brackets. Columns are then glued square into platform with *Araldite* or similar adhesive, but not glued to the lintels, (Figure 6.5, Photo 10). This last operation is left until the crankshaft, bearing shells and the flywheel are machined and ready for assembly.

5 BOTTOM TURNED COLUMNS

The four columns that support the engine platform are machined from $1\frac{5}{64}$+" = 1.088" (27.63mm) square pieces, 12" (304.8mm) long, with capitals and pedestals of the same square measurement $\frac{3}{4}$" (19mm) high. The wood is marked as follows: $\frac{25}{32}$+" = 0.787" (20mm) from each end left un-machined, and a length of $10\frac{27}{64}$" = 10.425" (264.8mm) for shaping as in Fig.6.6. Each end is centre-drilled and drilled $\frac{3}{16}$" or 5mm to a depth of $\frac{3}{16}$" to take a similar size dowel. The dowels should have a short projection for insertion into the underside of the platform, (Photo 11).

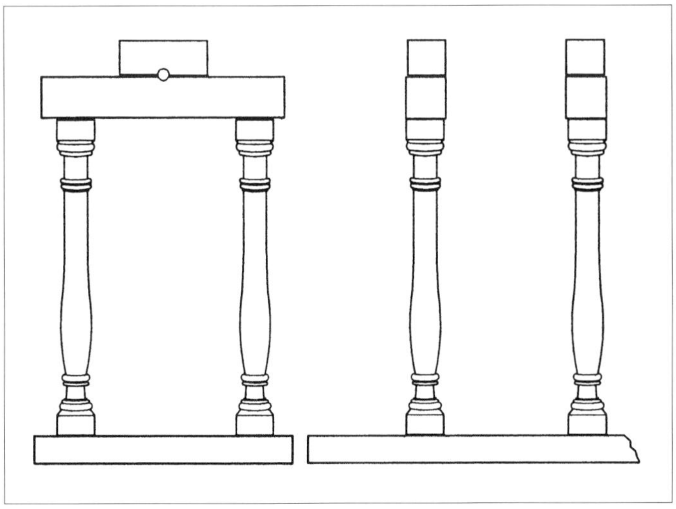

Fig. 6.5:
Sub-assembly - Stage 1. Front and side elevations of top structure

Photo 10: Sub-assembly of top structure

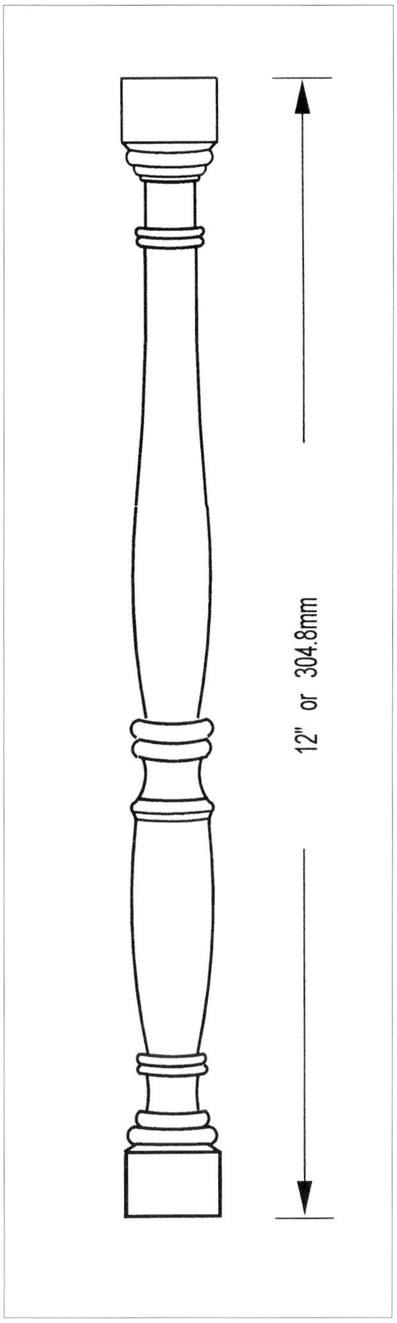

Fig. 6.6: Bottom turned column (x 4)

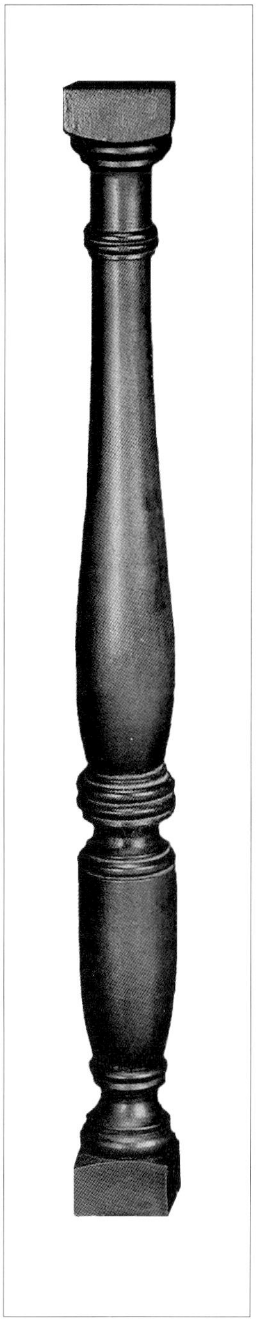

Photo 11: Bottom turned column (x 4)

6 BALUSTERS OR HORIZONTAL TURNED BARS

Two horizontal balusters are cut from 1" (25.4mm) square pieces, 9⅜" = 9.375" (238mm) long, with square ends of the same measurement and ¾" = 0.75" (19mm) long. The wood is marked 19mm from each end, leaving 7⅞" = 7.874" (200mm) for turning and shaping as in Figure 6.7, (Photo 12).

Both ends are drilled 3⁄16" or 5mm to a depth of 3⁄16" to take 7⁄16" = 0.437" or 11mm long dowels for fitting also into the base cross-bars.

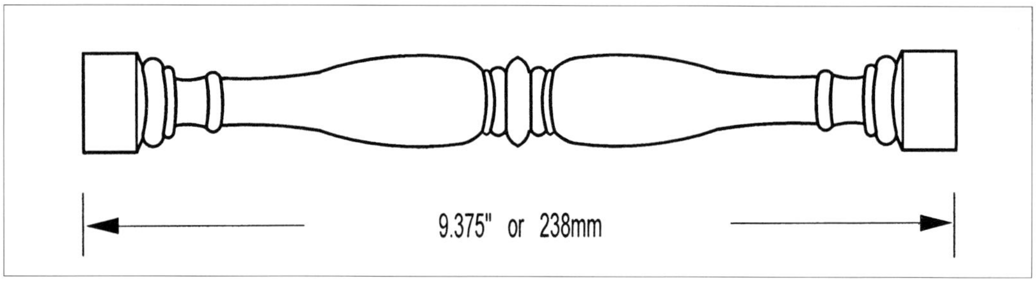

Fig. 6.7: Horizontal baluster (x 2)

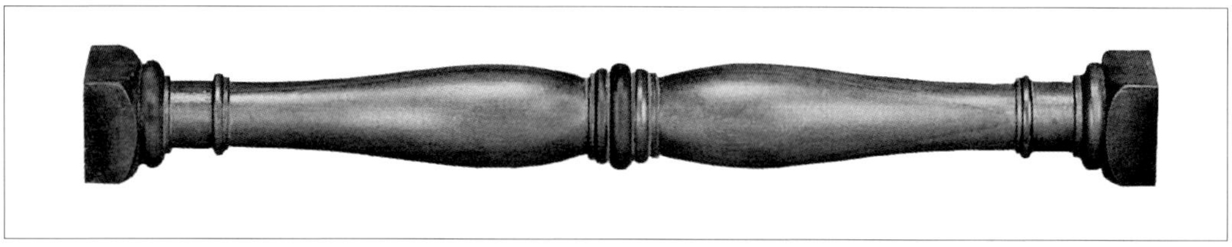

Photo 12: Baluster (x 2)

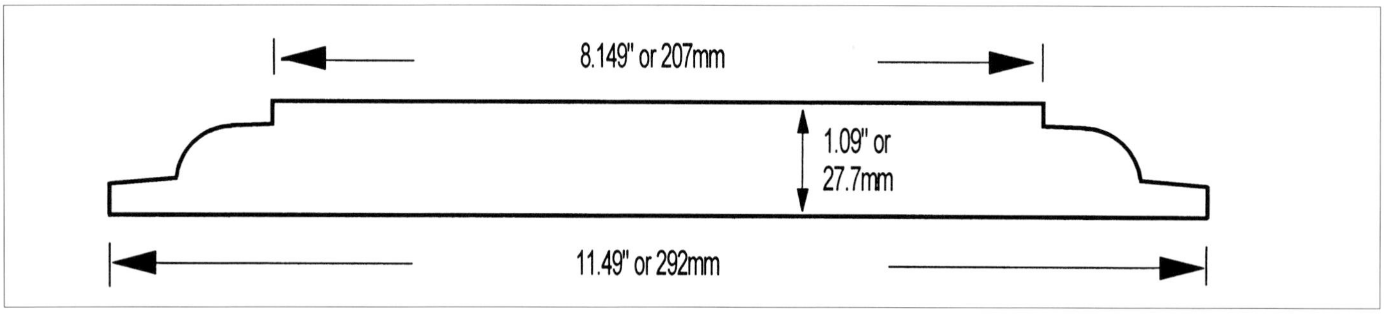

Fig. 6.8a: Profile of base cross-bar

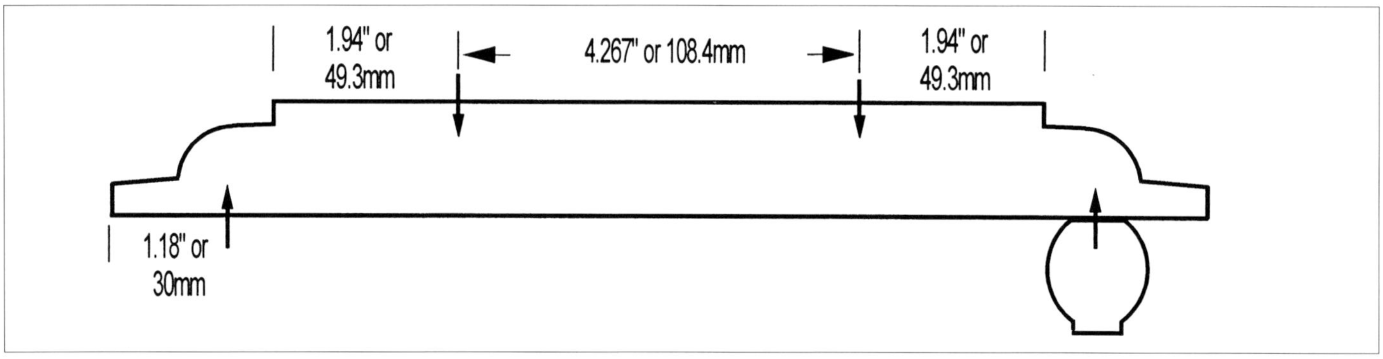

Fig. 6.8b: Marks for bottom columns and for rounded ball-feet

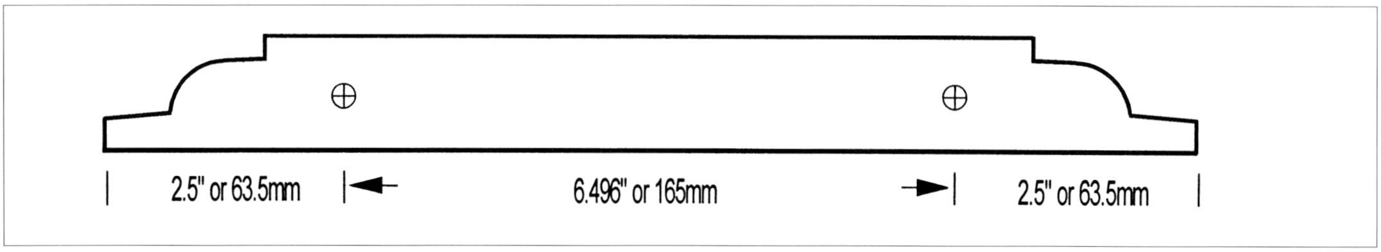

Fig. 6.8c: Marks for horizontal balusters

Fig. 6.8: Horizontal cross-bars (x 2)

7 BASE CROSS-BARS

Two base cross-bars that form the base of the model are cut 11½–" = 11.49" (292mm) long, 1³⁄₃₂+" = 1.102" (28mm) wide and 1³⁄₃₂" = 1.09" (27.7mm) high, shaped as in Figure 6.8a. They are marked on the top surface 1¹⁵⁄₁₆+" = 1.94" (49.3mm) from each end of the surface, 4¹⁷⁄₆₄" = 4.267" (108.4mm) between centres; drilled ³⁄₁₆" or 5mm for a depth of ³⁄₁₆" to take similar size dowels. These take the four bottom columns.

The cross-bars are turned over, the bottom marked 1³⁄₁₆–" = 1.18" (30mm) from each end and drilled ³⁄₁₆" to take dowels to a depth of ³⁄₁₆". These take the four semi-rounded ball-feet, O.D. 1¹⁄₁₆–" = 1.06" (27mm) and 1⅛–" = 1.122" (28.5mm) high with a projecting flat bottom. The ball-feet are similarly drilled to take the dowels. At this stage the ball-feet are glued and inserted into the base cross-bars (Figure 6.8b).

The cross-bars are then turned on their side so that when marked and drilled the inner side of the left hand cross-bar faces the inner side of the right hand cross-bar. Both are marked 2½" (63.5mm) from each end and drilled to take similar dowels for the balusters. (Figure 6.8c).

8 SUB-ASSEMBLY – STAGE 2

The balusters are fitted to the cross-bars, followed by the bottom turned columns which in turn are fitted into the engine platform (Figure 6.9). The top and bottom halves of the engine structure are then glued together as in Fig. 6.10, (Photo 13).

Photo 13: Sub-assembly of wooden structure of the Edinburgh model

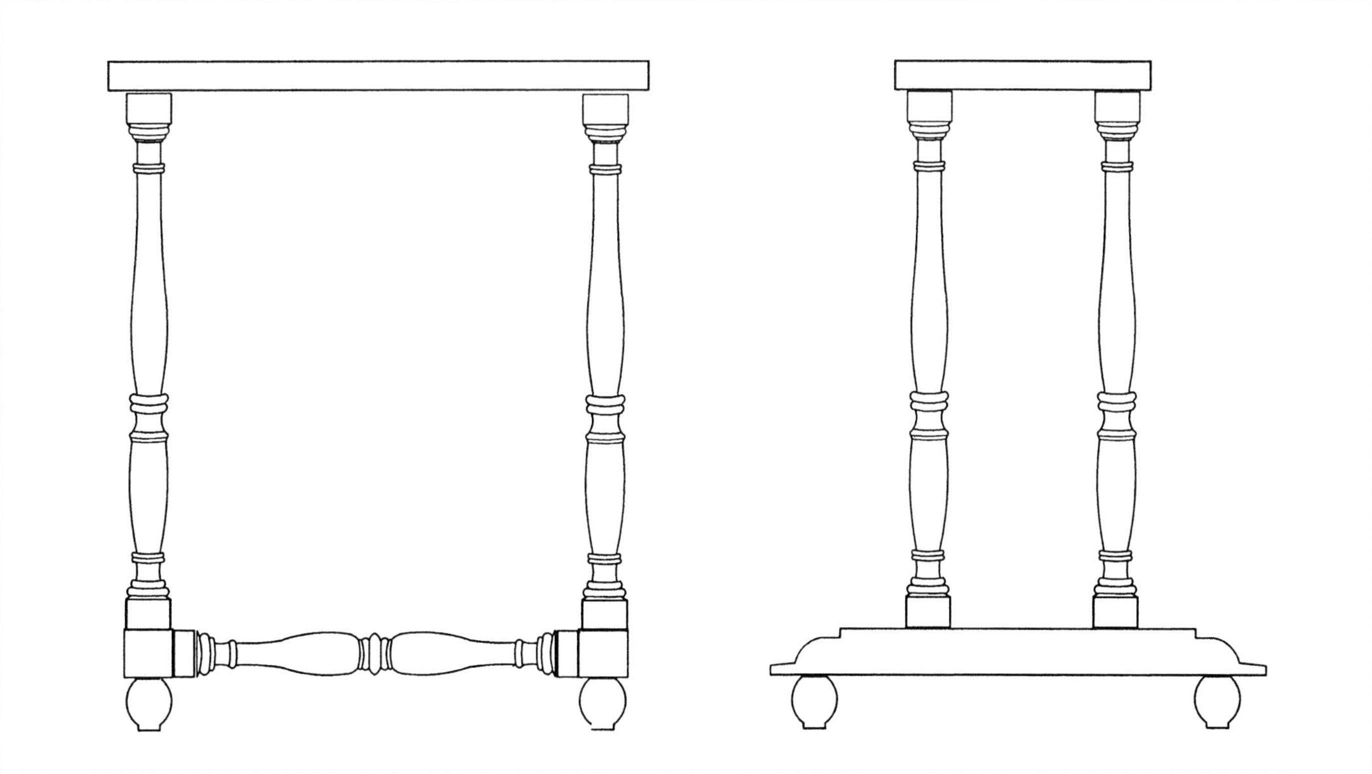

Fig. 6.9: Sub-assembly - Stage 2 Front and side elevation of bottom structure (¼ scale approx.)

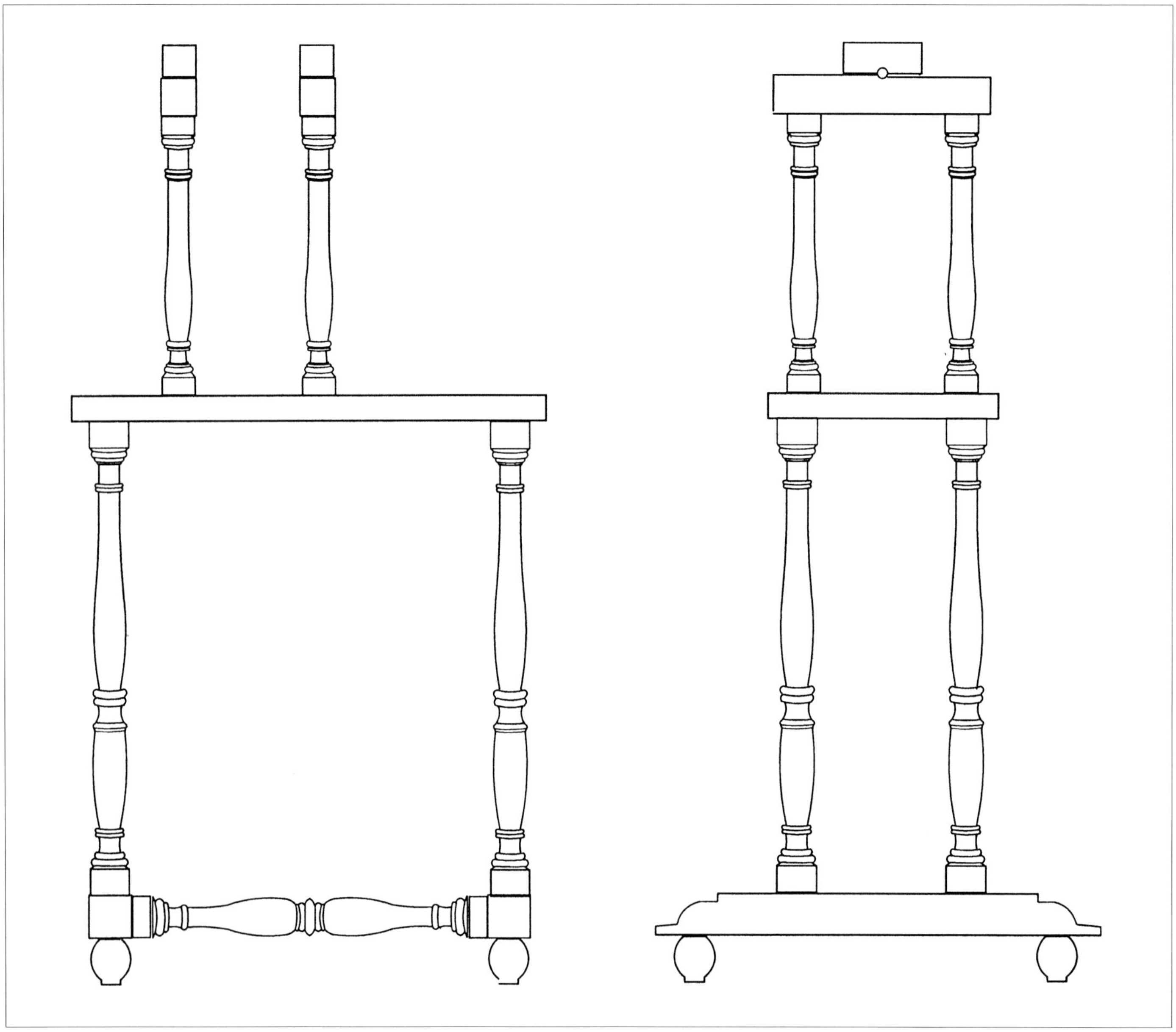

Fig. 6.10: Front and side elevation of completed wooden structure (1/4 scale approx.)

Construction Sequence 9 - 11

OTHER WOOD COMPONENTS

9 POWER CYLINDER PLATFORM

The power cylinder platform is in effect a distant piece to raise the level of the power cylinder to a height that allows a pressure transfer pipe from the displacer cylinder top to be soldered into the bottom of the power cylinder .

The platform is cut from mahogany $2\frac{17}{64}$" = 2.267" (57.6mm) wide (as seen from the front), $2\frac{53}{64}$+" = 2.834" (72mm) long (as seen from the side) and $1\frac{27}{64}$–" = 1.418" (36.04mm) high. Part of the surface of the platform is hollowed out to accommodate the pressure transfer pipe from the displacer cylinder which goes into the base of the power cylinder. The part that is hollowed is only ¾" = 0.75" (19mm) long and $\frac{15}{32}$" = 0.47" (12mm) deep. Finally the platform is drilled centrally to take a 4mm diameter screw (Figure 6.11).

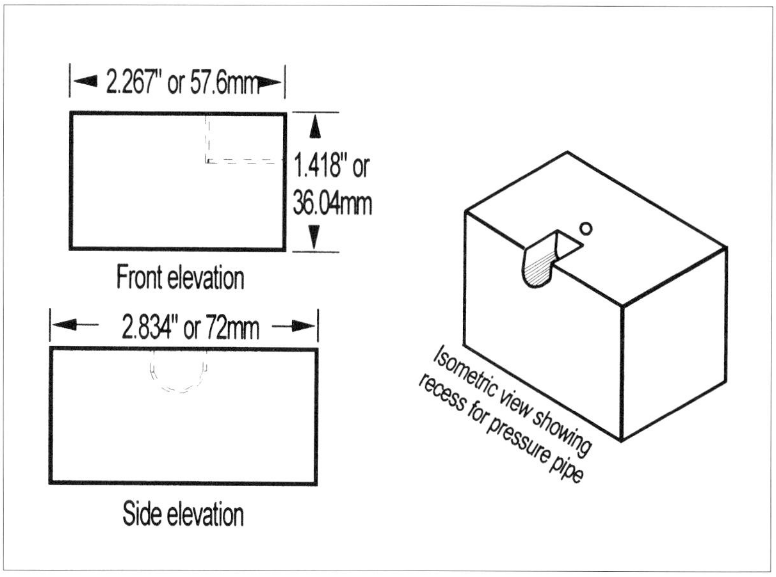

Fig. 6.11: Power cylinder platform

10 PULLEY

The pulley is machined from mahogany $\frac{31}{32}$" = 0.97" (24.63mm) thick, with an external diameter of $2\frac{13}{64}$" = 2.204" (56mm), drilled ⅜" = 0.375" (9.5 or 10mm). Before parting off, the pulley is reduced internally to $1\frac{57}{64}$" = 1.89" (48mm) internal diameter with the sides sloping inwards to give an open U-shape $\frac{39}{64}$+" = 0.612" (15.54mm) wide at the bottom, (Figure 6.12).

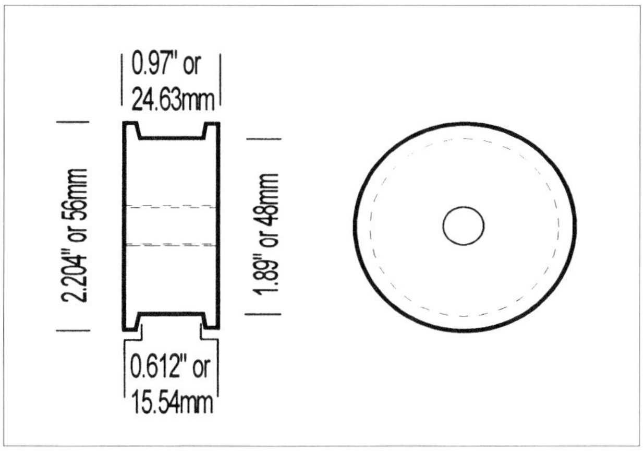

Fig. 6.12: Pulley

11 FILLET

The fillet is machined from mahogany $\frac{15}{16}$+" = 0.944" (24mm) diameter, $\frac{61}{64}$" = 0.954" (24.24mm) long, tapering to $\frac{45}{64}$+" = 0.708" (18mm) with a short step at the end $\frac{15}{32}$" = 0.47" (12mm) diameter, drilled 3/8" = 0.375" (9.5 or 10mm), and shaped as in Figure 6.13.

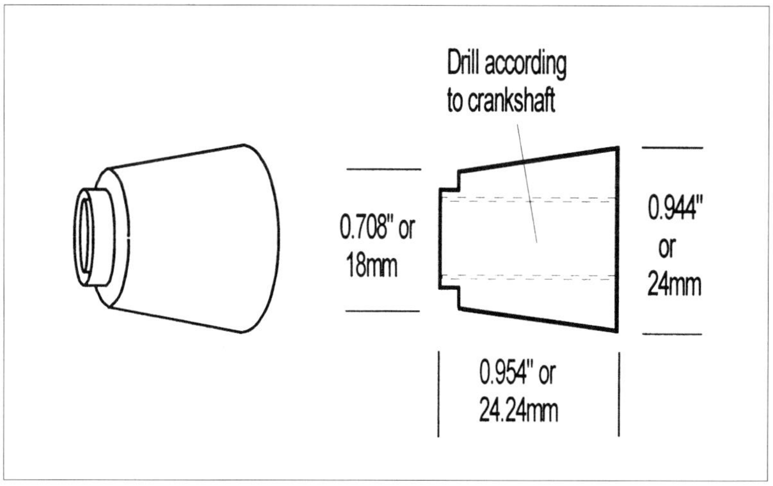

Fig. 6.13: Wooden fillet (R/H drawing - actual size)

Chapter 7
Metal Components - Edinburgh Model

Construction Sequence 12 – 16

NOTE: *While readers may prefer to keep as close as possible to historical materials used in Robert Stirling's models, there is no valid reason for refraining from using more modern approach with such components as cast iron pistons, brass bushes, aluminium alloy, etc..*

The original flywheel was made of cast iron to a high degree of finish that required little touching up or filing. Robert's brother James was a master craftsman in foundry in his early days of engineering and it is quite possible, even probable, that he had a hand in the casting of the components used in these models.

The flywheel and crankshaft of the Glasgow Hunterian Museum's model are cast in one piece, and it is therefore assumed that the Edinburgh model may have been similarly constructed. Unless one has access to a good pattern maker and foundry, it is difficult for the home technician to reproduce this type of composite component. The author has chosen to make the two parts, flywheel and crankshaft, as separate items.

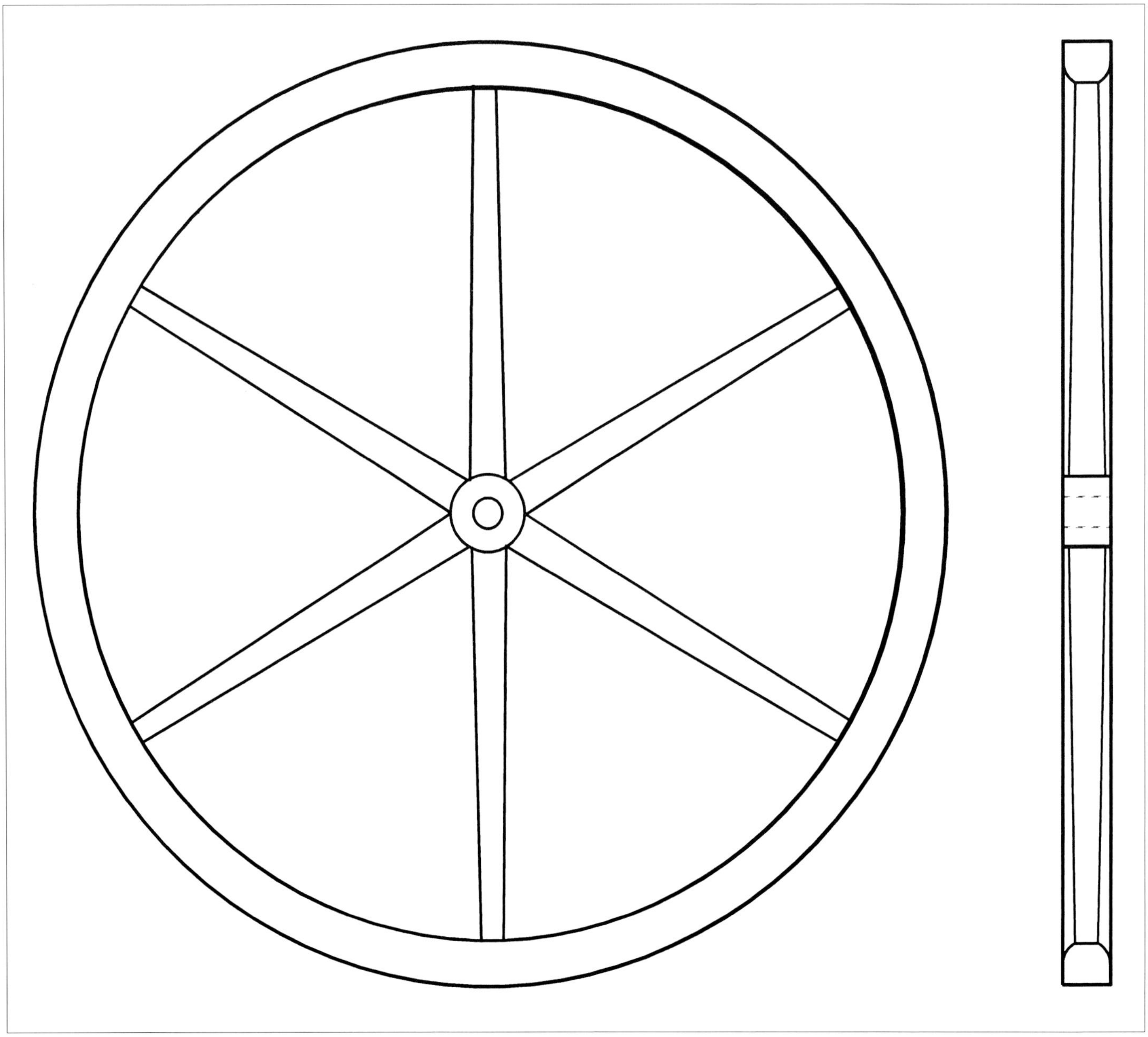

Fig. 7.1: Flywheel profile and side view

12 FLYWHEEL

The flywheel diameter is 12" or 304.8mm, rim thickness is 5/8" = 0.625" (15.875mm) with a flat top and rounded underneath to 9/16" = 0.562"(14.28mm), with six spokes approximately 5 25/64" = 5.393" (137mm) long to allow 15/64" = 0.234" (6mm) at each end for insertion into the rim and hub. The spoke thickness is 1/2" = 0.5" (12.7mm) at the centre tapering to 5/16" = 0.312" (7.935mm) at the rim. The hub is 1" (25.4mm) in diameter (Figure 7.1).

There are two ways to make a flywheel of this size:

1 make a wooden pattern cut from 3/4" (18-19mm) plywood, filed round and drilled in the centre, with the spokes reduced in thickness, have it cast in aluminium, then machine the circumference on a large lathe and finish it off either on a milling machine or hand-filing;

2 Draw the outline of the flywheel on a suitable size aluminium plate, drill a pilot hole in the centre for the crankshaft, chain-drill the outline of the circumference and the spokes, cut and file where necessary and finish off the circumference on a lathe or grinder.

{NOTE: In the 1/2 scale replica the 6" flywheel was made as follows: a similar size flywheel was cast in recycled aluminium, reduced in thickness and the original spokes/hub removed and replaced by wooden spokes made from 19/6–" = 0.295" (7.5mm) wood dowel, each 2 5/8–" = 2.618" (66.5mm) long to allow insertion into the rim and the hub, each dowel reduced in diameter to 1/4"(6.35mm) at the hub and tapering to 5/32" = 0.156" (4mm) at the rim end. Two spokes were inserted on opposite sides of the rim, cut to size to meet exactly in the centre and then glued together. Araldite was used in the rim, at the centre and in all the following stages of the operation. Later two spokes were cut to size to meet in the centre and glued to the first pair. The process was repeated with the last pair (Photo 14).

Photo 14: Pattern for making a 6" flywheel casting

The whole assembly was left overnight, then carefully centre-drilled and drilled 5/32" = 0.156" or 4mm. Two pieces of aluminium bar, O.D. 1/2" and 1/2" long shaped as in the original model, were drilled 5/32" or 4mm, glued and bolted on opposite sides of the spokes. The end result was good enough to serve as the actual flywheel for the replica, however it was decided to use it as a pattern for two castings that could be used on both the Edinburgh and Glasgow 1/2 scale engines (Photo 15a & 15b). The flywheel was then drilled through 3/16" = 0.187" (4.75mm)}

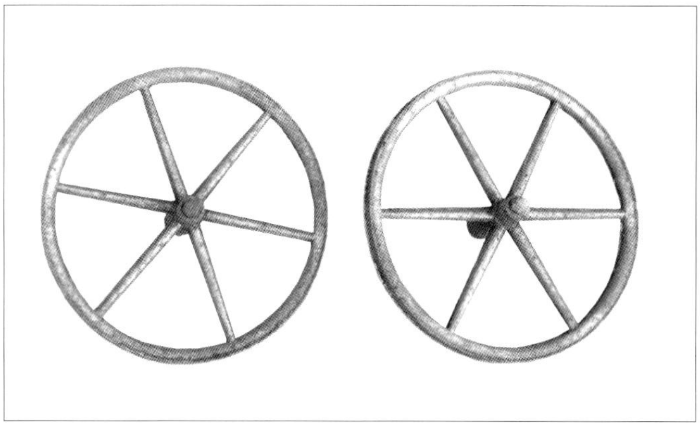

Photo 15a: Castings made from pattern

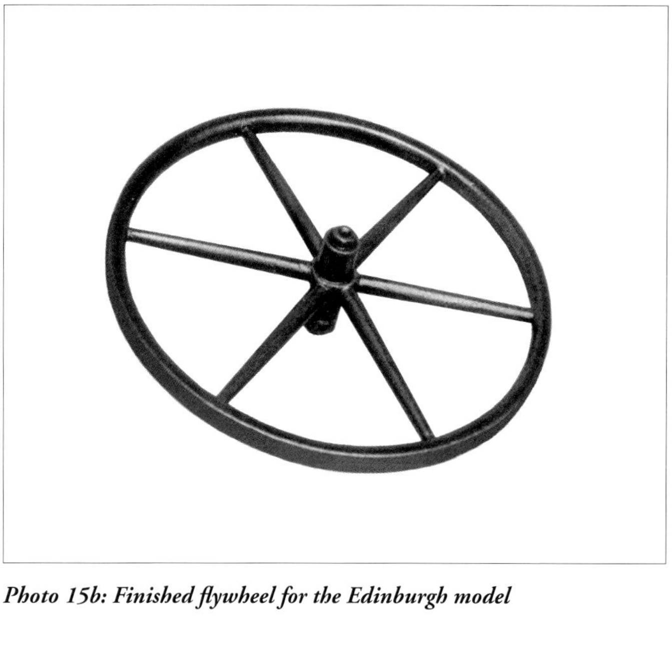

Photo 15b: Finished flywheel for the Edinburgh model

13 CRANKSHAFT

The crankshaft total length is 6 19/64+" = 6.3" (160mm). It tapers at each end towards the cranks from 3/8" = 0.375" (9.5 or 10mm) at the fillet/pulley centre to 7/32+" = 0.228" (5.8mm), while the centre part, 2 19/32" = 2.6" (66mm) long is left at 3/8". This centre part of the crankshaft takes the flywheel, pulley and fillet (Figure 7.2a).

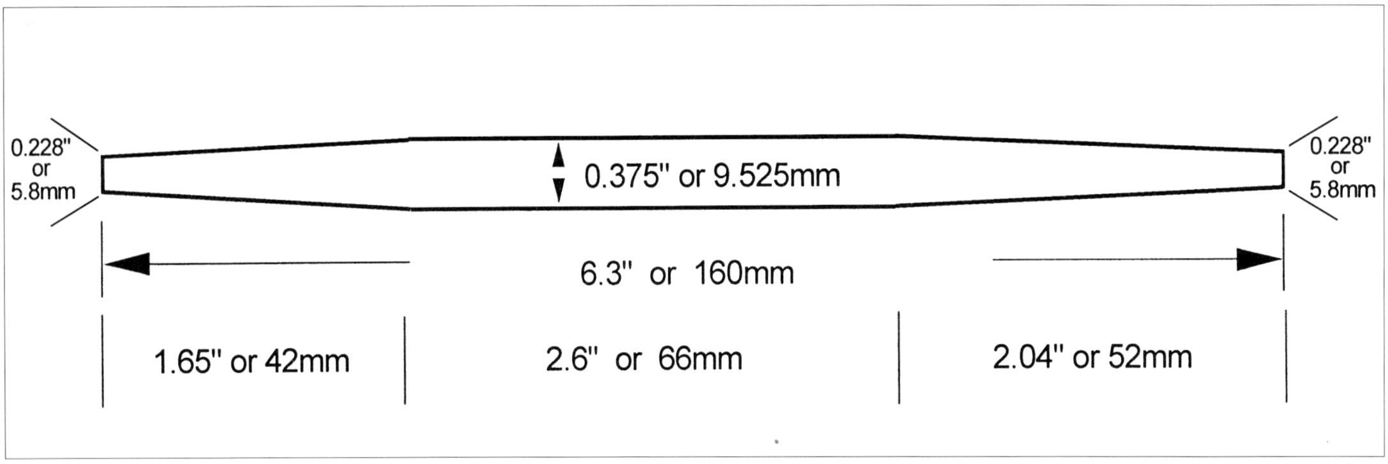

Fig. 7.2a: Crankshaft (Actual size)

At the power cylinder end the crankshaft is cross-drilled with a 1/16" or 1.5mm drill to secure the crank with a pin. (The displacer end is treated differently with the crank having a 6BA securing bolt – this because, it is thought, the phase angle had still to be experimented on.)

Bearing Shells

(NOTE: Robert Stirling elected to insert bearing shells between the lintels and the securing brackets for the crankshaft to rotate with minimum friction. From images of the model it appears that these shells were made from thin lead or copper sheets, pressed around the tapering ends of the crankshaft. Both the top and bottom shells had a wing-like extension on either side of the crankshaft end so that the shells could be retained in place by the pressure of the top securing brackets.)

The bearing shells are cut from thin copper or lead sheet in two parts, pressed in a vice around the ends of the crankshaft, trimmed and inserted between the lintels and the upper brackets after these were filed down to accommodate the shells tight fit. (Figure 7.2b).

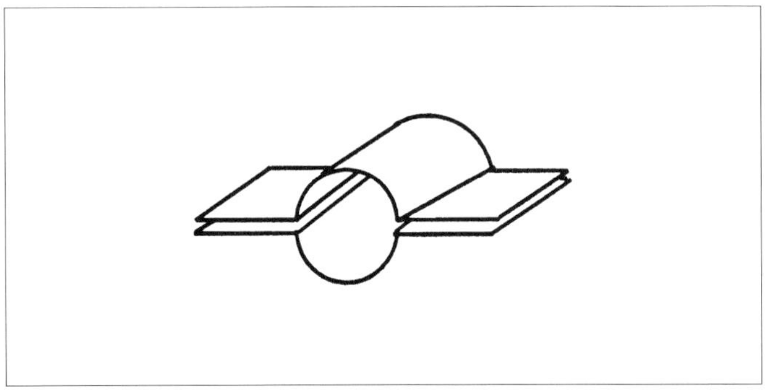

Fig. 7.2b: Bearing shells (Enlarged)

14 SUB-ASSEMBLY – Stage 3

The crankshaft is inserted tight into the flywheel with the boss and pulley on either side leaving a projection of 1 41/64+" = 1.65" (42mm) on the power side and leaving a projection of 2 1/32+" = 2.04" (52mm) on the displacer side (Figure 7.3).

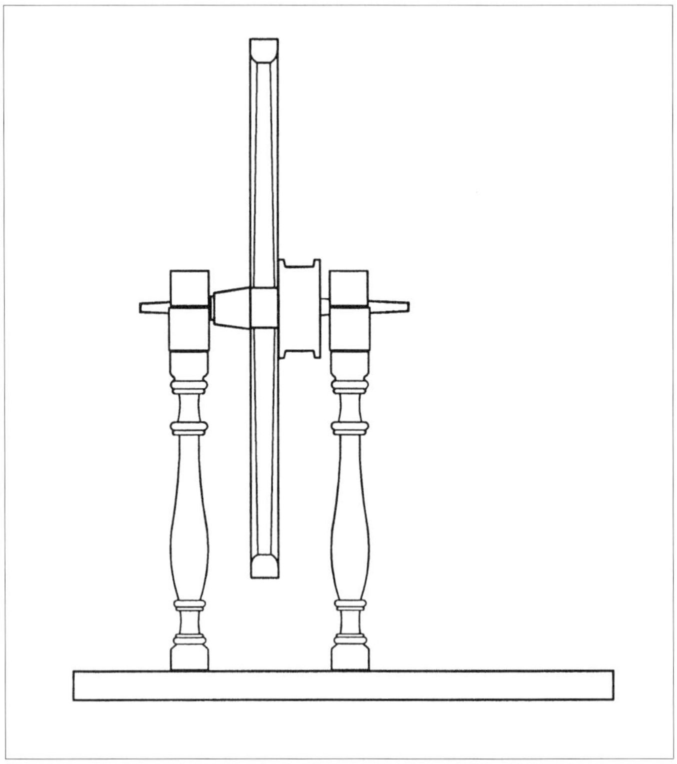

Fig. 7.3: Sub-assembly - Stage 3 Flywheel, pulley and fillet on crankshaft, supported by top engine frame (1/4 scale)

The crankshaft assembly is placed between the bearing shells on the lintels and secured by the short brackets, each with two countersunk 4mm x 30mm screws. A trial turning by hand should indicate if some minor filing is required in the area where the tapered crankshaft sits. Obviously, even with some lubrication, the result cannot not be as good as with brass/bronze bearings or ball races, (Photo 16).

Photo 16: Sub-assembly with flywheel on top structure with temporary Crankshaft

15 POWER CYLINDER, PRESSURE TRANSFER PIPE, PISTON, CLEVIS, CON-ROD AND FITTINGS, POWER CRANK AND PIN

POWER CYLINDER

The specifications for the full size cylinder are as follows:

External Length:	3⅛+"	=	3.137"	79.7mm
Internal Length:	3 1/64+"	=	3.02"	76.7mm
O.D.:	2 1/16–"	=	2.06"	52.32mm
I.D.:	1 61/64"	=	1.95"	49.62mm
Base flange:	2 11/64+"	=	2.18"	55.42mm x 3mm

The cylinder is hollowed out of solid brass, faced externally to the above outside diameter (52.32mm) leaving a flange-type projection O.D. 55.42mm, 3mm thick at the bottom. The cylinder is then bored, honed and lapped to give an internal diameter of 1 61/64" = 1.95" (49.62mm). (Figure 7.4) Photo 17

The cylinder is then drilled in the centre to take a 4mm countersunk scew, and drilled between the centre and the inner wall to take a short pressure pipe O.D. 15/32" = 0.47" (12mm).

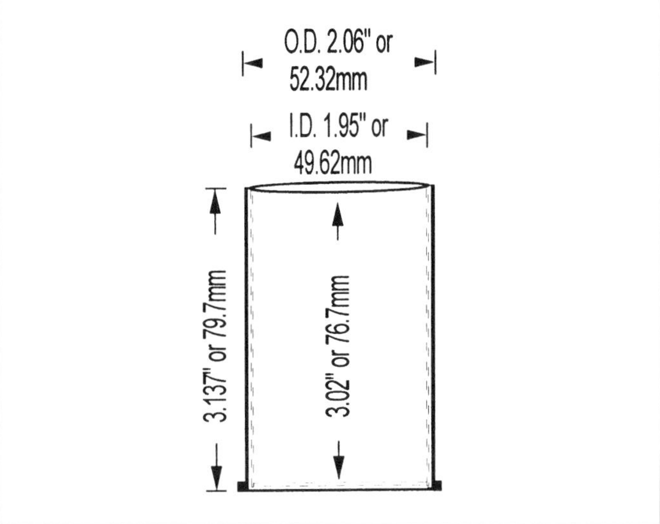

Fig. 7.4: Power cylinder

Photo 17 : Power cylinder

PRESSURE PIPE CONNECTING SYSTEM

NOTE: ***The pressure transfer pipe and its connections are best first described holistically and then the parts apportioned to the respective cylinders. Stirling elected to make the connection between the two cylinders with a number of components, and therefore it is important to see and study the whole set-up first, and go into detail about the construction of each individual part after. The description below deals with the whole connection first.***

The connections can be divided into three sections, the pipe connected to the power cylinder, the pipe connected to the cylinder head of the displacer cylinder and the centre section which brings the two together. A threaded socket was soldered to the displacer cylinder pipe and therefore this is considered to be part of the pipe, Figure 7.5. The pipe from the power cylinder is long enough to almost reach the displacer cylinder. The two pieces that bring the two cylinders together are a sleeve and a connector. In effect the sleeve becomes part of the connector in the way it is mounted with a ridge or lip that fits securely in a recess in the connector. The connector in turn was internally threaded with the same thread as the socket on the displacer pipe.

What happened on assembly, therefore, was that before the power cylinder was bolted to the base, the sleeve/connector fitting was inserted on the pipe from the power cylinder, but allowed to slide until its position vis-à-vis the displacer cylinder pipe was established correctly.

Once the power unit was in place the connector was screwed on the displacer pipe socket and then the sleeve soldered to power pipe. In this way the connector brought the two pipes together with an air-tight joint. From personal observation of the connector, and the socket at displacer end, it could be seen that all the parts were well machined and the threads very finely cut, no mean achievement 180 years ago.

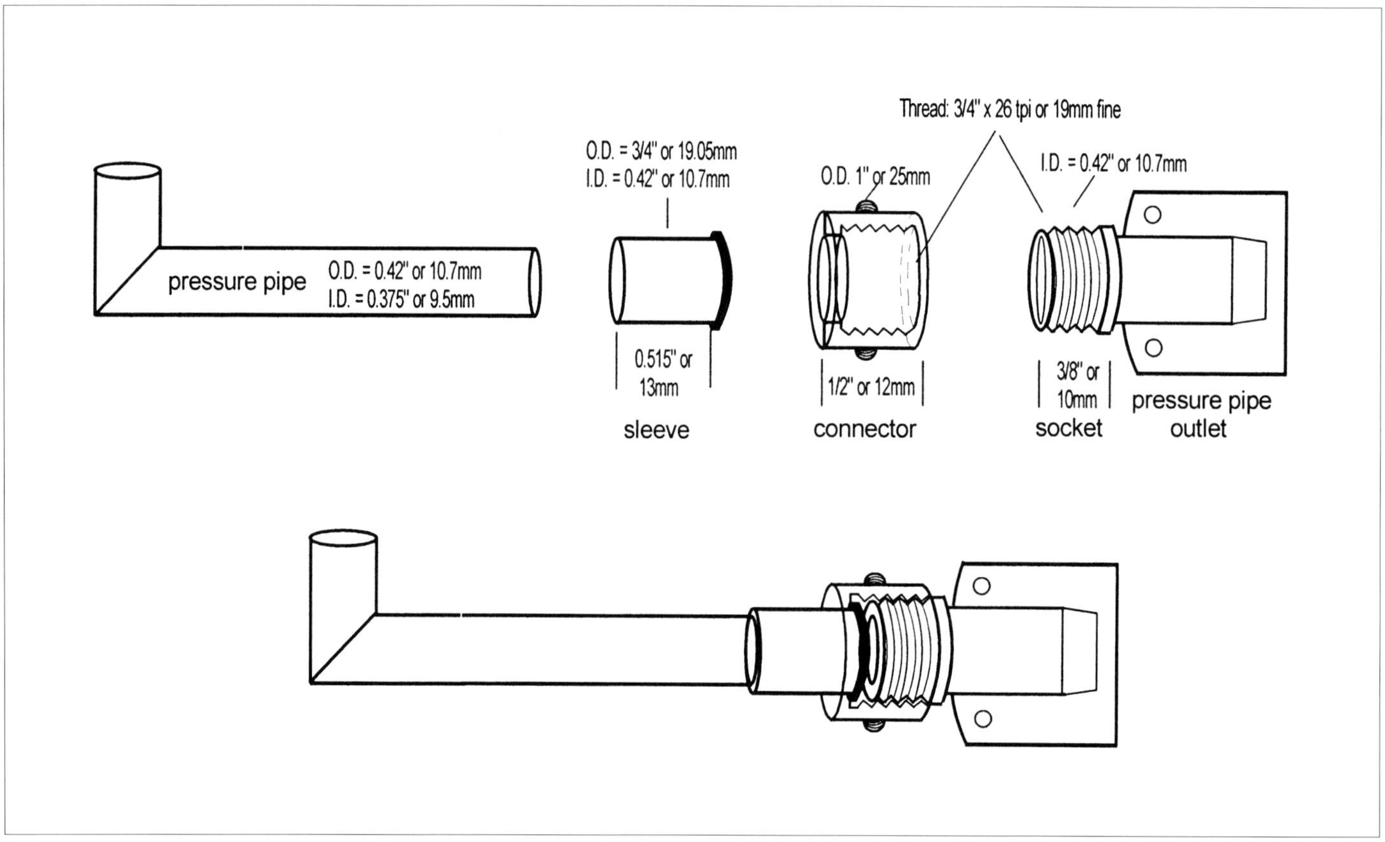

Fig. 7.5: Layout of the pressure pipe connecting system between the power and displacer cylinders (Not to scale)

Power cylinder pressure pipe

The pressure pipe that enters into the base of the power cylinder is cut from thin walled brass pipe $4\frac{23}{32}$" = 4.72" (120mm) long, O.D. $\frac{27}{64}$" = 0.42" (10.7mm), I.D. $\frac{3}{8}$" = 0.375" (9.5mm).

The pipe is marked, cut at 90° and soldered as in Fig. 7.6a and 7.6b. At this stage the fitting into the power cylinder base is only temporary until the pipe end-connector and sleeve are machined and the length of pipe confirmed (Fig. 7.6b). Figure 7.6c shows the power cylinder on the wooden platform with a profile of the pressure pipe.

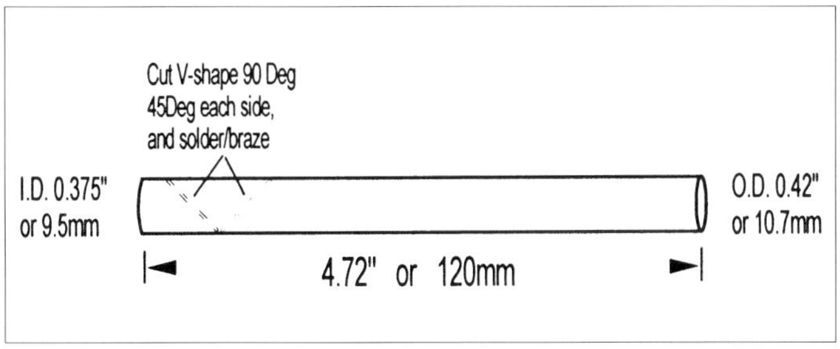

Fig. 7.6a: Pipe prepared for bending at 90°

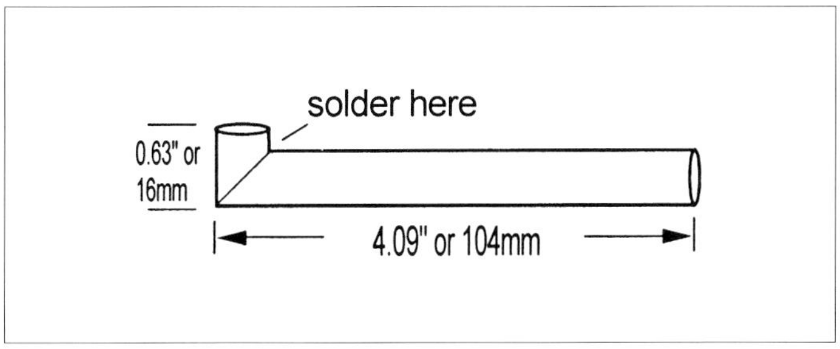

Fig. 7.6b: Pipe bend soldered

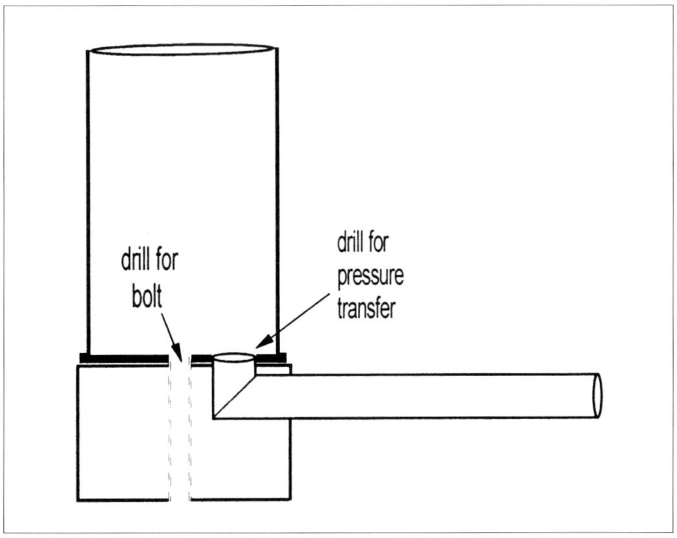

Fig. 7.6c: Pipe soldered to cylinder base

A sleeve is machined or cut from brass pipe $\frac{19}{32}$" = 0.59" or 15mm long, O.D. $\frac{3}{4}$" = 0.75" or 19.05mm, I.D. $\frac{27}{64}$" = 0.42" or 10.7mm. The sleeve is faced and reduced in external diameter to $\frac{1}{2}$" or 12mm for a length of $\frac{33}{64}$" = 0.515" or 13mm, leaving a 2mm ridge or lip (18mm O.D.) finished square. The sleeve has a tight fit both on the pressure pipe from the power cylinder and into the connector with a perfect seating.

The connector is a small fitting machined from solid brass rod O.D. 1" or 25mm and $\frac{1}{2}$" or 12mm long, drilled through $\frac{15}{32}$" = 0.47" or 12mm to take the sleeve body in a tight fit. The hole is then widened to I.D. $\frac{45}{64}$" = 0.703" (18mm) for a length of $\frac{5}{16}$" or 0.312" (8mm) and then threaded $\frac{3}{4}$" = 0.75" (19mm) × 26 t.p.i. or similar fine thread, with an undercut at the end. The end is finished square. This is the part that takes the sleeve in an airtight fit. The connector is slightly reduced at each end to $\frac{7}{8}$" = 0.875" (22mm) leaving a collar in the centre which is then knurled at conform to the original model, (see Figure 7.21).

The sleeve and connector are mounted on the pressure pipe from the power cylinder head, but no further work is involved at this stage.

POWER PISTON

(NOTE: ***In the early days of hot air/caloric engines the power piston was usually covered with soft leather either greased or soaked with light lubricating oil. Leather-covered pistons had the advantage to conform with any slight irregularity of the cylinder diameter. As such these pistons were quite efficient and some engines worked for very long periods before requiring a change of leather. Readers may opt to construct the piston in the same manner (to keep to the historical aspect) or to machine a cast iron one to give a gas tight fit in the honed cylinder.***

Although we do not know for sure if the Edinburgh Museum engine leather piston is double-sided there are two indicators that this is so. The cut-away replica of the Edinburgh model held at the Science Museum, Kensington shows clearly that the piston is double-sided and since this model is an exact reproduction there leaves little room for doubt. Similarly it is confirmed that the Glasgow Hunterian Museum's model has a working piston made up of 'two greased leather pads folded at the edges and sandwiched between two wooden discs – held on to the connecting rod by a brass nut' – description by Mike Jewkes of the Hunterian Museum. The author has assumed that the Edinburgh Museum model has a similar type of double-padded leather piston.)

Photo 18: Aluminium jig and method of making leather pistons (Courtesy - Roy Darlington and Keith Strong)

A leather-covered piston, while requiring less precision than a gas-tight fitting cast iron one has to move or glide easily along the cylinder with minimum lubrication since this can cause unnecessary friction. The leather has to be very thin, light in weight, absorbent and malleable.

A detailed description of a leather covered piston can be found in "STIRLING and HOT AIR ENGINES", by Roy Darlington and Keith Strong (Pages 117 – 119). The description given here is a short adaptation of the process described by Roy. Composite Photo 18 is reproduced from Roy and Keith's book with their kind permission.

"The first step requires that the leather be worked for malleability by soaking in water for at least 24 hours, then squeezing the leather and rubbing it between fingers and stretching it over a wooden mandrel until a cup is formed. A wooden mandrel with a slightly rounded end is prepared, narrower than the internal diameter of the cylinder by the thickness of the leather when squeezed dry. If for example the leather is $^1/_{16}$" = 0.0625" or 1.6mm in thickness, and the I.D. of the cylinder is $1^{61}/_{64}$" – 1.953" (49.62mm), the diameter of the mandrel is between $1^{25}/_{32}$+" to $1^{53}/_{64}$" = 1.79" to 1.826" (45.5 – 46.4mm). Actually a better idea is that the mandrel is machined even narrower than mentioned above so that the leather can form a cup-like shape in stages. Once the first stage has been achieved, the diameter of the mandrel is increased in steps by winding a piece of electrical tape. The success of a leather-covered piston lies in the final shape of the leather cup which must not have any creases.

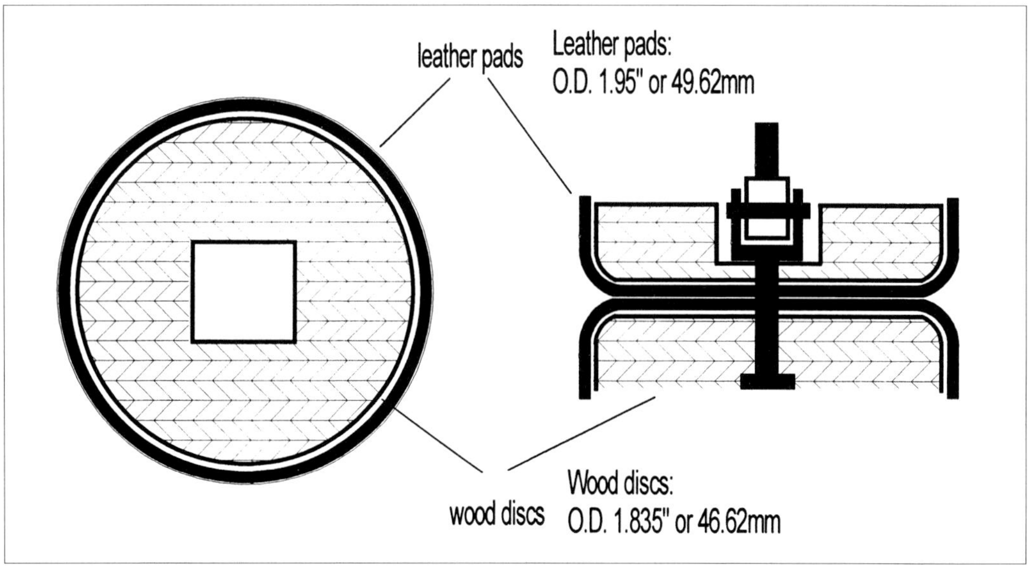

Fig. 7.7: Power piston with 2 leather pads between 2 wooden discs with clevis-type device holding the two parts together (Actual size)

A jig is machined from aluminium bar OD $2\frac{23}{64}$" = 2.362" (60mm), internally bored $1\frac{61}{64}$+" = 1.95" (49.62mm), $\frac{15}{32}$" = 0.47" (12mm) deep, slightly rounded in the base to the shape shown in Figure 7.7 The wet leather is dressed over the mandrel and slowly inserted into the jig under slight pressure for a few millimeters at a time without creasing it. When fully inserted and left to dry, the excess leather is cut with a scalpel or a very sharp hobby knife. The process is repeated for the second pad."

In the meantime two hard wood discs (beech or oak) are prepared, OD $1\frac{53}{64}$+" = 1.835" (46.62mm)**, $\frac{25}{64}$" = 0.393" (10mm) thick, rounded as in Figure 7.7. The bottom one is centre-drilled and drilled $\frac{5}{32}$" = 0.157" (4mm) and recessed slightly to take the bolt head. The top disc requires more machining: first drilled as for the bottom disc, the hole is then widened into a ½" or 12mm square shape and $\frac{5}{16}$" = 0.312" (8mm) deep.

*(**Note: ... only if the leather is 1.5mm or slightly less in thickness. Always adapt diameter of discs according to the thickness of the leather).*

CLEVIS and CON-ROD SMALL-END

A clevis-type device (Figure 7.8) is machined to bolt the two discs together; the component is made up of two parts, the bottom part is a 4mm bolt with a straight shank and threaded 8BA or 2mm for a length of $\frac{3}{32}$+" = 0.098" or 2.5mm. The top part is the actual clevis and is machined from ⅜" = 0.375" or 10mm brass cube, milled ¼" or 6mm across and $\frac{19}{64}$" = 0.296" or 7.5mm deep, cross-drilled to take a $\frac{1}{16}$" or 1.5mm pin. The centre of the hole is either 4.5mm from the bottom or 3mm from the top. This part is then drilled and tapped 8 BA or 2mm to take the bottom bolt-type fitting.

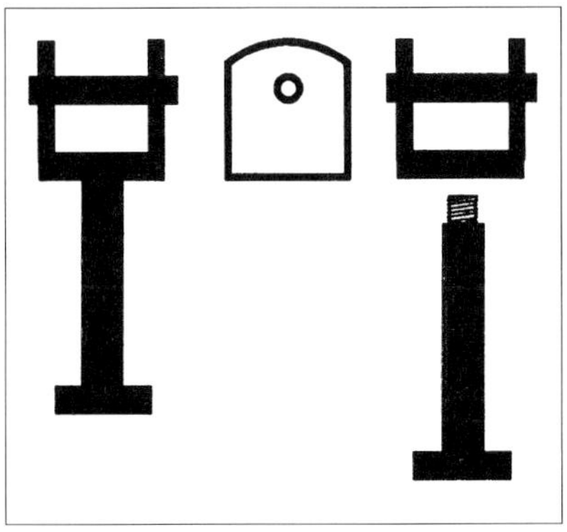

Fig. 7.8: Clevis device (Actual size)

The con-rod small-end fitting that goes into the wooden disc is machined from $\frac{5}{16}$" = 0.312" (or 8mm) brass rod, ¼" or 6mm thick, centre-drilled and drilled $\frac{1}{16}$" or 1.5mm in the centre and cross-drilled and threaded 6 BA or 3mm to take the con-rod end (Figure 7.9). The small-end is filed down on both sides to just under ¼" or 6mm in thickness so that it can turn easily in the clevis.

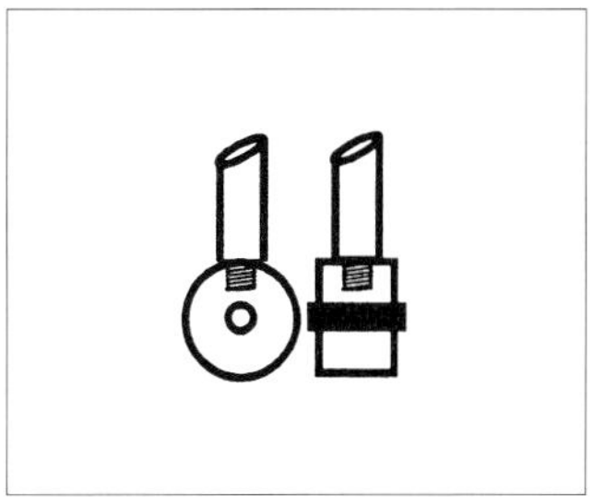

Fig. 7.9: Con-rod small-end (Actual size)

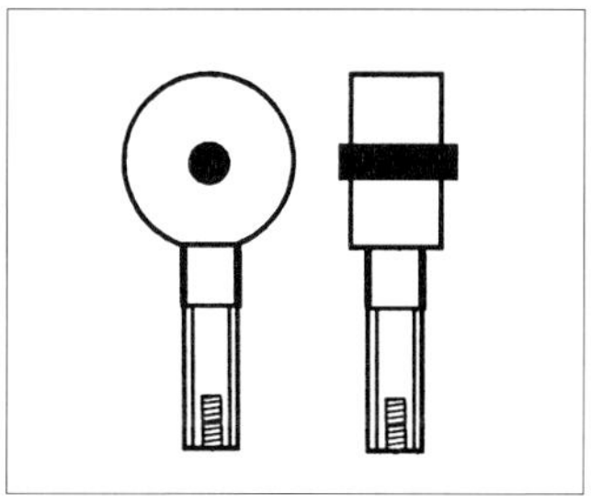

Fig. 7.10: Con-rod big-end (Actual size)

Finally the leather pads are dressed over the wooden discs held by the clevis device and inserted with a temporary con-rod into the working/power cylinder with a few drops of light oil and worked up and down manually until the leather pads bed down.

CON-ROD BIG-END

The big-end is machined from $\frac{3}{16}$–" = 0.18" (4.57mm) thick brass, total length 1" (25.4mm) long, in three sections, a $\frac{31}{64}$" = 0.48" (12.2mm) round end, a narrow square neck $\frac{5}{32}$" = 0.156" (4mm) wide and $\frac{9}{64}$–" = 0.134" (3.4mm) long, ending in a hexagon of *almost* the same diameter and $\frac{25}{64}$" = 0.39" or 10mm long. The top round end is centrally drilled to take a $\frac{5}{32}$–" = 0.151" (3.85mm) crankpin, while the hexagon part is carefully drilled and tapped 6BA or 3mm to a depth of $\frac{5}{32}$" = 0.156" or 4mm. (Figure 7.10).

POWER CON-ROD

(NOTE: *the following applies to this section and all the following sections where connecting rods are described. Where a connecting rod has two threaded ends, and the length between the centre points of two end-components is subject to alteration – plus or minus – it is always advisable to invest in Left Hand and Right Hand taps and dies (the most common in imperial range are 2 BA, 4 BA and 6 BA – metric according to availability). Threading the con-rod ends with R/H and L/H dies, and tapping the receiving components with identical taps – R/H to R/H etc., allows the machinist to make fine adjustment to the length of the con-rod without the need to take the con-rod away from its position on the crank and in the clevis.*

The power con-rod is cut from ⅛"= 0.125" or 3.2mm or $\frac{9}{64}$" = 0.140" or 3.5mm mild steel rod approximately 3¾" = 3.74" (95mm) long, threaded both ends 6BA or 3mm. When assembled with the two end components the distance from mid centre of the big-end to mid-centre of the small-end is $4\frac{29}{64}$" = 4.45" (113mm), (Figure 7.11). *Readers should however check the actual length of the rod when the crankpin is at its lowest point and the power piston just above the base of the power cylinder. Any fine adjustment in the length of the con-rod can be done with screwing in or out the ends.*

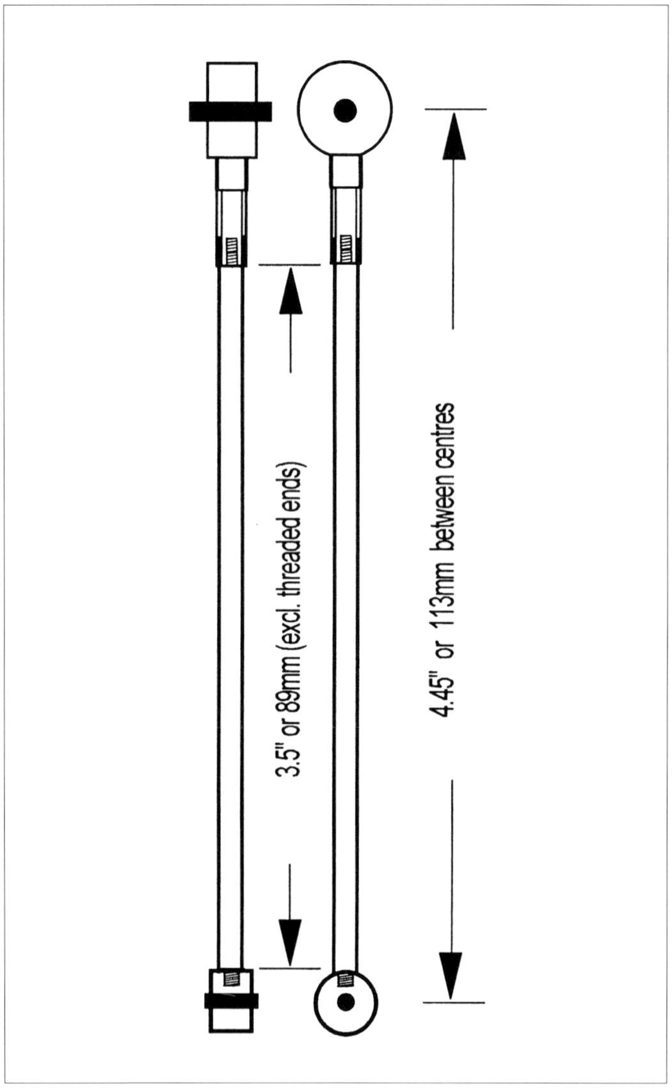

Fig. 7.11: Power con-rod

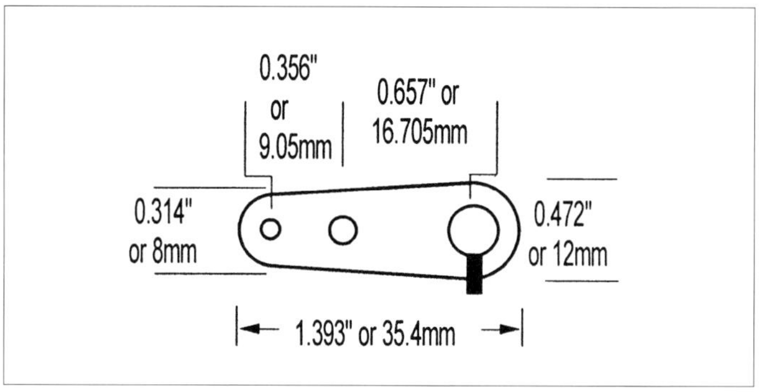

Fig. 7.12: Power crank

Three holes are drilled in the crank. Three marks are made as in Figure 7.12. At the wide end the drilled hole tapers from 7⁄32+" = 0.228" (5.8mm) to 15⁄64" = 0.234" (5.95mm) as in Figure 9.13. A second hole is drilled and tapped 4BA or 3mm at a distance 21⁄32" = 0.657" (16.705mm) from the centre of the crankshaft hole and the third hole drilled and tapped as the second, at a distance of 1+" = 1.014" (25.755mm) from mid-centre of the crankshaft hole. The crank is then cross-drilled with a 1⁄16" or 1.5mm drill at the point where it is to be mounted on the crankshaft (Figure 7.13). *(In Stirling's model the crank was subsequently pinned to the crankshaft end.)*

The crankpin is machined from mild steel rod, reduced to 5⁄32–" = 0.151" or 3.85mm O.D., 11⁄16+" = 0.694" (17.64mm) long, threaded 4BA or 3mm for a length of 15⁄64" = 0.234" or 6mm. At the other end the crankpin is drilled to take a retaining pin to hold the con-rod big-end in place. The crankpin is then screwed into the crank (inset in Fig. 7.13).

POWER CRANK and CRANKPIN

(NOTE: An interesting feature of this crank (and of the displacer crank) is the fact that Robert Stirling wanted to experiment with different strokes for both the power piston and the displacer presumably in order to find the optimum performance of the engine. To obtain different strokes of the power piston, Stirling opted to drill and tap two holes away from the crankshaft end, one at a distance of 21⁄32" = 0.657" (16.705mm) and the second at a distance of 1+" = 1. 014" (25.76mm) from the centre of the crankshaft end, thus giving two alternative strokes to the piston, one of 15⁄16+" = 1.315" (33.41mm) and a longer one of 2+" (51.52mm). In the end the crankpin was inserted in the first tapped hole and a shorter stroke was achieved.)

The power crank is 125⁄64" – 1.393" (35.4mm) long, 13⁄64-" – 0.200" (5.08mm) thick at the crankshaft end, tapering to 5⁄32+" – 0.163" (4.15mm) at the bottom end, 15⁄32+" – 0.472" (12mm) wide at the top tapering to 5⁄16" – 0.314" or 8mm before rounding off.

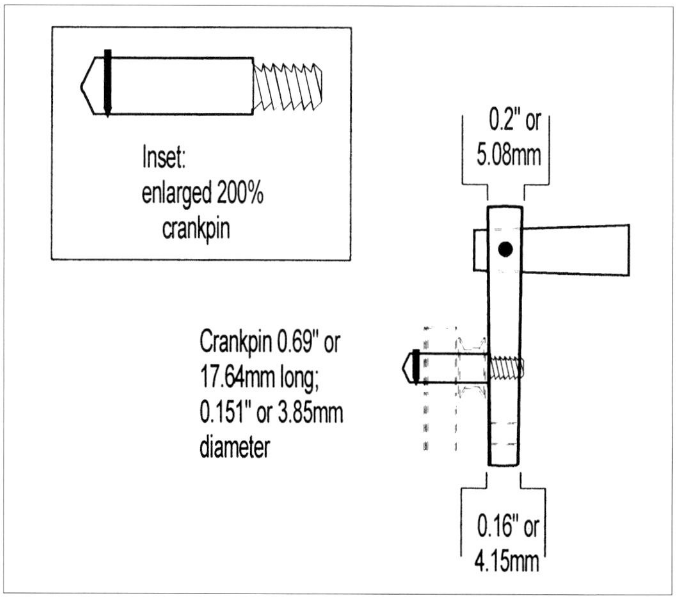

Fig. 7.13: Power crank with crankshaft end, crankpin and positions of con-rod big-end and fillet

Photo 19: Sub-assembly with power unit coupled to the crankshaft

16 SUB-ASSEMBLY – STAGE 4

(Temporary assembly) The power cylinder and wooden base (Photo 19) are temporarily screwed to the engine platform, and the power piston/con-rod inserted into the cylinder. The crank is mounted on the crankshaft and the con-rod big-end inserted on the crankpin and secured lightly with the retaining pin. A few turns of the flywheel will show whether both the stroke and the piston fitting work well. In fact if the pressure transfer pipe hole is stopped by a finger, the piston should not budge. (Figure 7.14).

Construction Sequence 17 – 20

DISPLACER UNIT

17 DISPLACER CYLINDER

(NOTE: Most of the metal parts in Stirling's Edinburgh model were made of brass or copper with the exception of the displacer. Normally we do not nowadays use brass/copper for the displacer cylinder because of the heat conductivity along its length which makes the engine liable to slow down as the temperature between the hot and cold ends lessens considerably after a short run. In the ½ scale replica brass tubing was used partly in preference to copper, but it also so happened that a 38mm OD brass pipe, with an ID of 36mm was readily available at the local metal stockist.

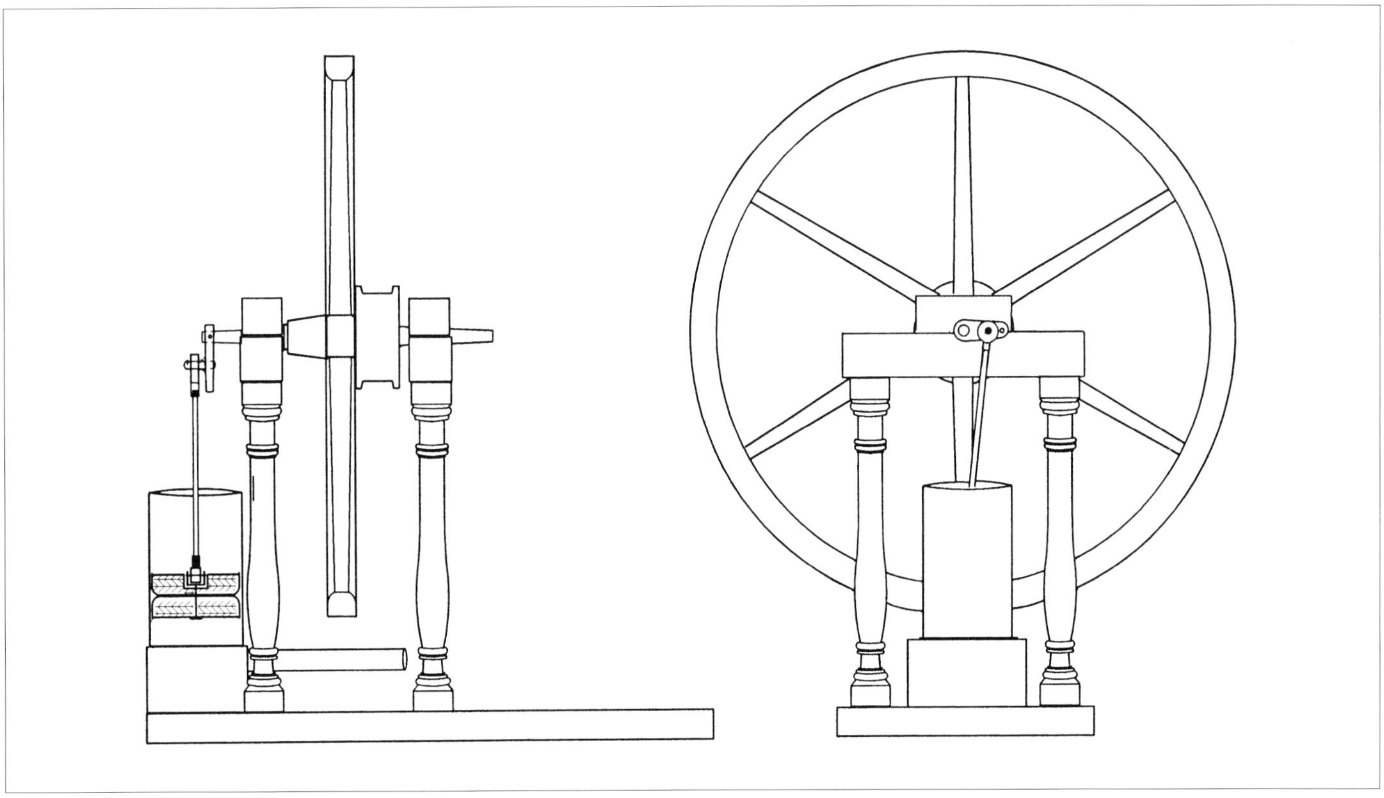

Fig. 7.14: Sub-assembly - Stage 4 Power unit on engine platform and attached to crankshaft

Photo 20: An old photograph of the displacer cylinder and displacer of Edinburgh model

Stirling's Edinburgh model appears to have a vertical seam – probably soldered – which indicates that the cylinder was made from rolled copper sheet. The sealing disc was a beaten hollow shell which overlapped the cylinder and also soldered. Photo 20 shows an old photo of the displacer unit when it was dismantled many years ago for some repairs. Photo 21 shows the cylinder's vertical seam and a horizontal seam about ⅓rd of its length from the base in what appears to have been an extension to the cylinder.

Photo 21: A close-up view of the cylinder displacer showing the vertical and horizontal seams

The displacer cylinder is machined from brass tube (or bright mild steel) O.D. 2³¹⁄₃₂" = 2.97" (75.52mm), I.D. 2⅞" = 2.87" (73mm) and 11¹⁄₁₆" = 11.063" (281mm) long. A disc ¹⁄₁₆" = 0.625" (1.58 – 1.6mm) thick and slightly larger in diameter than the cylinder is brazed to the base leaving a small extension to conform with the original model, (Figure 7.15)

A flange is machined from brass with the following measurements: OD 3⁹⁄₃₂+" = 3.285" (83.46mm), while the internal diameter is that of the cylinder's external diameter, and ³⁄₁₆+" = 0.197" or 5mm thick. The flange is then marked in two places for securing brackets to be soldered on the cylinder.

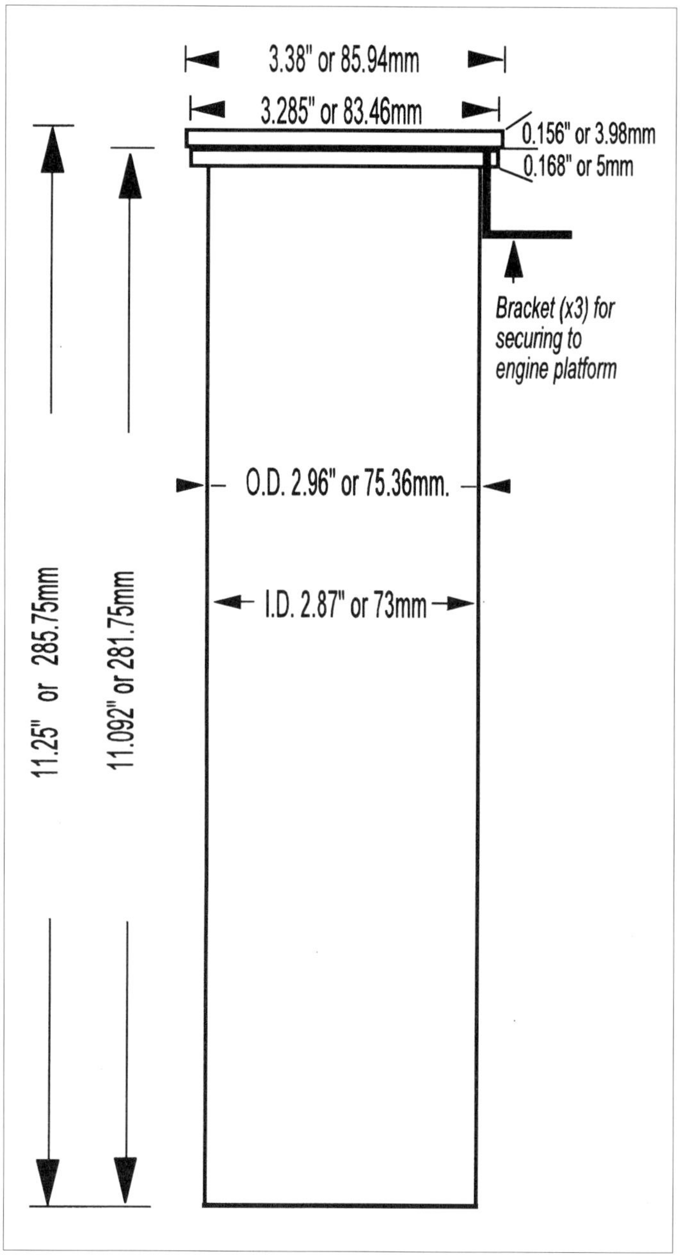

Fig. 7.15: Displacer cylinder

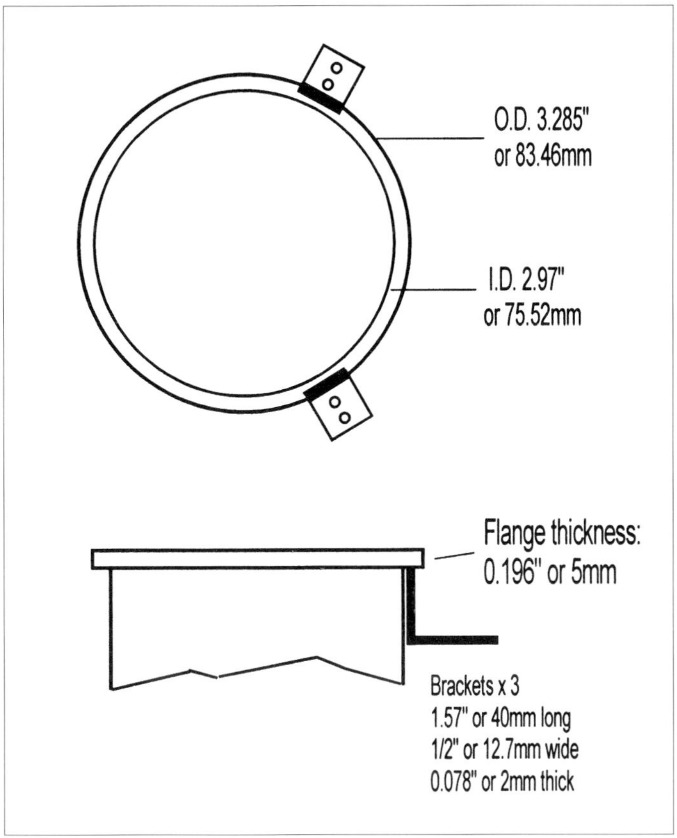

Fig. 7.16: Displacer cylinder flange

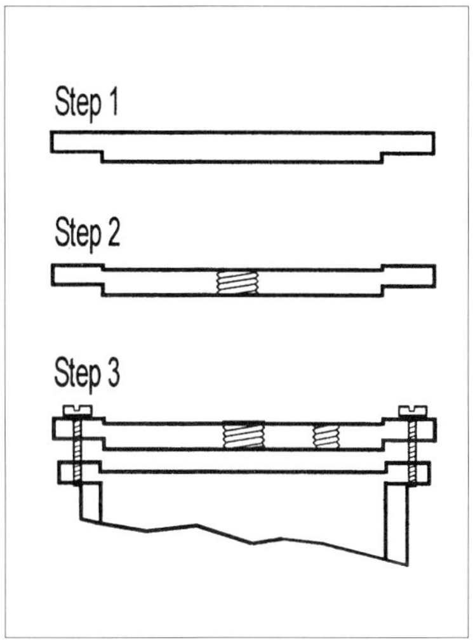

Fig. 7.17a: Side elevation of cylinder head during machining and assembly

Each bracket is cut from 5⁄64" = 0.078" (2mm) thick brass strip, 19⁄16+" = 1.574" (40mm) long and ½" (12.7mm) wide, bent at 90°, the vertical arm being ¾" (19mm) long, while the horizontal arm is 13⁄16" = 0.812" (21mm approx.) long and drilled to take two 4mm x 8mm long brass screws. The flange and brackets are soldered to the cylinder. At the appropriate time these brackets are screwed to the engine platform to hold the displacer cylinder in place, (Figure 7.16).

CYLINDER HEAD

(The sealing flange or cylinder head together with the pressure pipe outlet can easily be made as one casting if the machinist has access to casting facilities. The author elected to do this in preference to two separate components in his scaled down model. Photo 22 shows two separate components made with French chalk which were used to make a bronze casting of the complete cylinder head – actual size.)

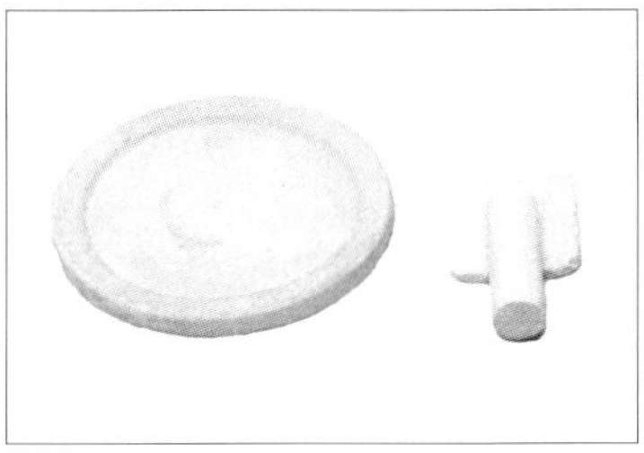

Photo 22: Pattern for the cylinder head made with French chalk

The machining of the cylinder head can be made with reference to Figures 7.17a and 7.17b which give both the machining process and the profile of the components. The cylinder head is machined from brass with a diameter of 3⅜" = 3.38" (85.94mm), 15⁄64" = 0.236" (6mm) of which 5⁄64" = 0.079" (2.02mm) thickness fits exactly into the cylinder while 5⁄32" = 0.156" (3.98mm) projects over the top of the displacer flange (Step 1).

The top is given a shallow recess on the top surface to conform with the original model (Step 2). While in the chuck the cylinder head is centre-drilled, drilled, and tapped 19⁄32" = 0.59" or 15mm ** to take the displacer rod guide bush. (Photo 23).

Photo 23: Finished casting of the displacer cylinder head with pressure pipe fitting

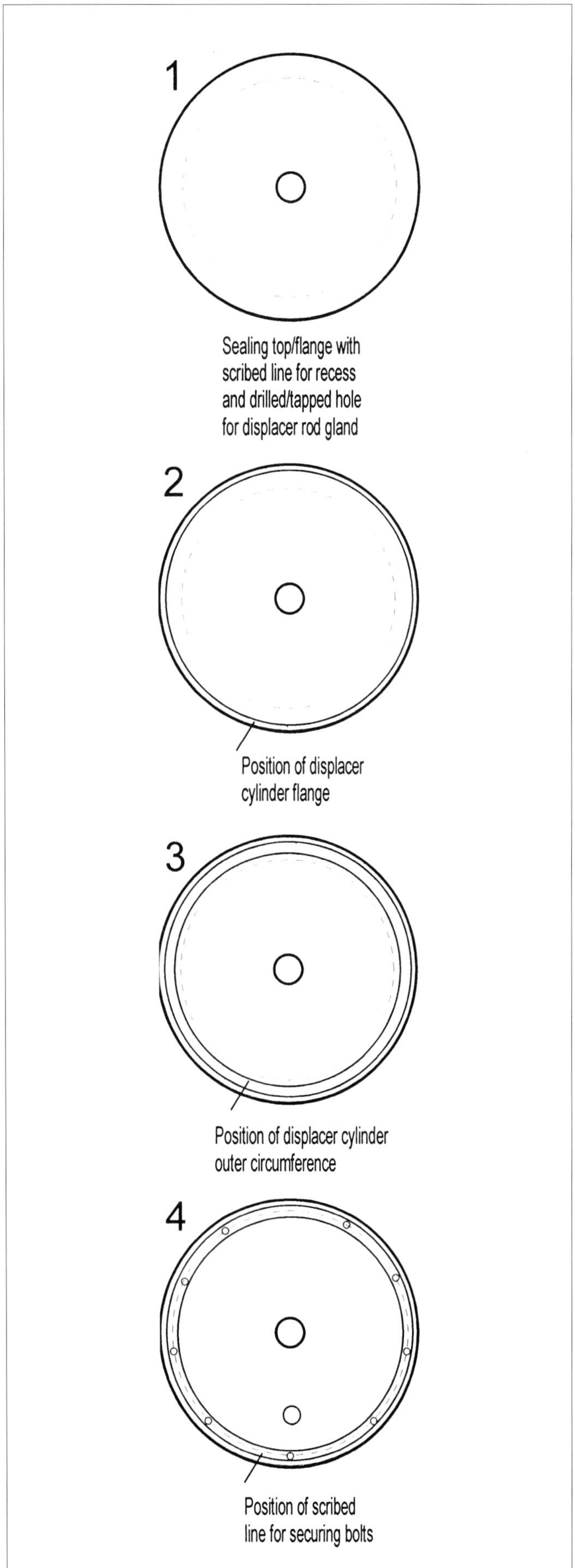

Fig. 7.17b: Profile of the cylinder head in stages of machining

****Before proceeding with this threaded part readers are directed to read 'machining note' in the section on the gland/ guide bush below.**

The cylinder head is reversed in the chuck and a line scored in the outer circumference so that 9 almost equidistant holes (taking into consideration the position of the pressure outlet) can be marked and later drilled to take 8 BA or 2mm bolts to secure it to the cylinder flange (Step3). The final job is drilling 5/16" = 0.312" (8mm) hole under the pressure transfer outlet and 13/64–" = 0.20" (5.1mm) hole on the opposite side for a pressure release vent which takes an 0 BA or 6mm threaded screw with an 'O' ring (See Figure 7.20).

(Note: the actual drilling for the 8mm hole under the pressure transfer outlet is done after the component is soldered on.)

DISPLACER ROD GLAND/GUIDE BUSH

(NOTE: The displacer rod gland or guide bush is one of the more important components in the displacer unit, since while it acts as a guide to the displacer rod, it has to be machined and assembled in such a way as to prevent any escape of air from this end. In Stirling's model the gland was machined in two parts with oil-soaked twine or thin leather in between.)

Machining notes:

1 ***this component has two identical threaded ends, 15mm actual. Reader may use B.S. Fine 9/16" or U.N.F. Fine 9/16"- the metric equivalent of 15mm is not difficult to find and/ or use;***

2 ***The displacer con-rod and displacer rod measured on Stirling's model vary from 3.2mm (1/8") to 3.43mm in diameter and are not quite uniform throughout their length, therefore reader may safely replace this either by a 1/8" (3.2mm) or a 3.5mm mild steel rod.***

The gland is machined into two parts in a number of operations, best described with reference to Figure 7.18 which shows the completed and assembled component on the left hand side and the two separate parts in the centre.

A brass rod 25/32" = 0.787" (20mm) diameter, at least 2½" or 63mm long is first faced and reduced to 23/32+" = 0.728" (18.5mm) for a length of 7/8–" = 0.871" (22.13mm), drilled 1/8"= 0.124" or 3.2mm (alternatively 3.5mm), then parted off. This is the bottom section that goes into the displacer cylinder head. The piece is placed in the chuck, drilled and tapped 15mm or equivalent for a length of 25/64–" = 0.387" (9.84mm). The piece is then reversed in the chuck, reduced and also threaded 15mm or equivalent for a length of 17/64+" = 0. 275" (7mm) - *(the actual length of thread required is 6mm, however a 1.5mm 'O' ring or greased twine on the threaded end ensures a gas-tight fit.)*

The top section of the gland is machined from remaining part of the rod first reduced to OD ¾+" = 0.757" (19.25mm) for a length of ¾+" = 0.761" (19.34mm) then drilled 1/8" or 3.5mm as above. There are three operations to be made on this component: a threaded end, a knurled part and an octagon shaped top.

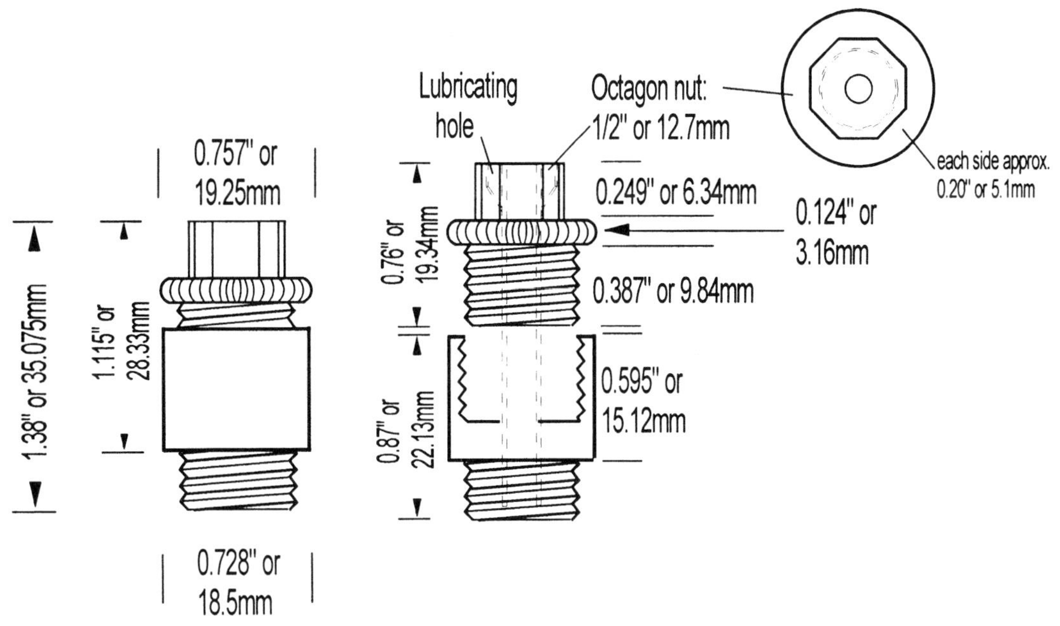

Fig. 7.18: Displacer gland with specifications (Approximate size)

The author suggests that the threaded end and the knurled part are machined first, and then the remainder is reduced to $\frac{19}{32}$" = 0.59" (15mm) diameter and later bench-worked and filed down to an octagon shape, ½" or 12.7mm across, each side of the octagon being approximately $\frac{13}{64}$–" = 0.20" (5.1mm).

The threaded end is ⅜+" = 0.387" or 9.84mm long, with the same thread as for the bottom component, i.e. B.S. Fine $\frac{9}{16}$" or U.N.F. $\frac{9}{16}$" or 15mm. The knurled part has a thickness of ⅛" = 0.124" (3.16mm) and is first slightly rounded then knurled to the shape of the original model (Fig. 9.18)

One last operation on the this component is to widen the bore on the octagon nut end into a shallow hollow ⅜" = 0.375" or 10mm deep (as it appears in images of this component), presumably so that oil could be placed inside to lubricate the displacer rod.

Finally the two parts are screwed into each other re-drilled and reamed to ensure perfect alignment.

PRESSURE PIPE OUTLET

***(See note in 'Cylinder Head' above** - The pressure pipe outlet in the Edinburgh model appears to be a cast bronze item, possibly even part of the casting of the cylinder head.)*

The pressure pipe outlet that connects the displacer cylinder to the power cylinder can be made either as a casting or fabricated from brass pipe and sheet, cut and soldered together in a way that provides an air tight junction. This is by far one of the more difficult parts to duplicate.

The construction of this outlet is best explained by reference to the drawings in Figure 7.19. The composite drawing shows the six steps it takes to construct the complete part. A piece of $\frac{27}{64}$" = 0.42" (10.7mm) OD thin walled brass pipe, 1$\frac{37}{64}$" = 1.574" (40mm) is cut first. The second step is to cut through half the pipe horizontally for 1" (25.4mm). Step 3 is to cut the back of the pipe at a sloping angle (to conform with the shape of the original model).

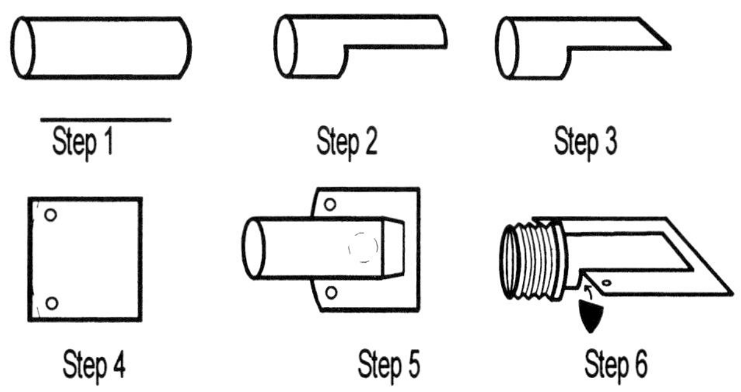

Fig. 7.19: Machining stages of pressure outlet

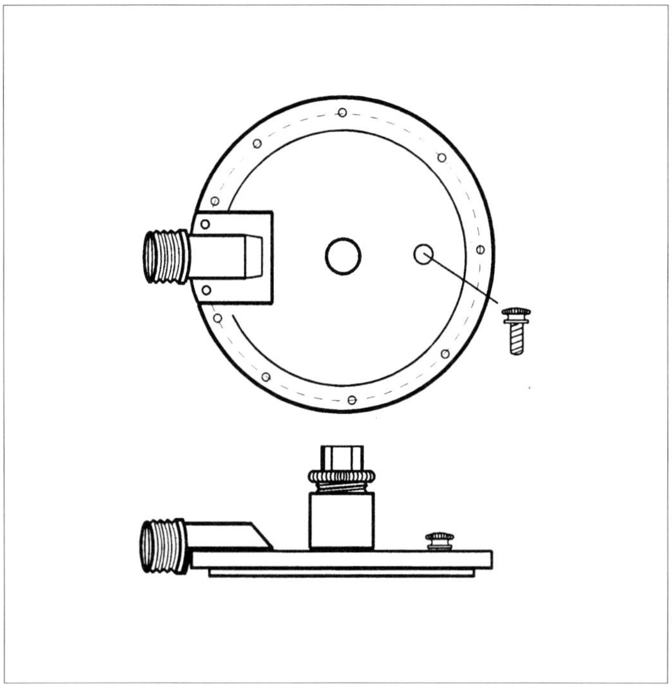

Fig. 7.20: Position of pressure pipe outlet on cylinder head and of vent screw

In step 4 a piece of brass sheet 5/64" = 0.0787" (2mm) thick is cut with the following dimensions: 1" (25.4mm) wide and 1 7/64–" = 1.102" (28mm) long. In step 5 the pipe is soldered to the brass piece along the sides, at the angled back and at the back of the pipe. **At this point this outlet is checked for any leakage from the joints.** The brass sheet base is filed round to conform to the diameter of the cylinder head, placed on the cylinder head and marked for drilling to take securing screws.

In step 6 a threaded socket is soldered to the pipe outlet. This takes the connector from the power cylinder pressure pipe (see Figures 7.5, 7.20 and 7.21).

The cylinder head is marked at the place where the outlet is to be soldered and the top rim is then filed flush with the surface of the recessed area so that the outlet can be soldered to the top without any problem of leakage. Particular care needs to be taken to seal the rear part of the pipe that almost touches the rim. A half-round piece of brass sheet is cut slightly larger than the diameter of the pipe and soldered to seal this end.

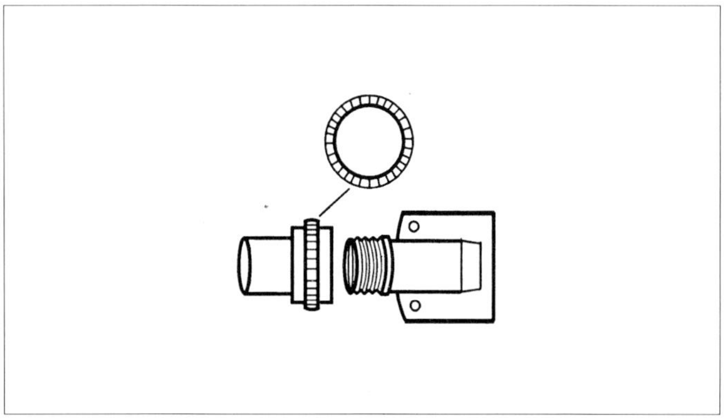

Fig. 7.21: Pressure connector and sleeve

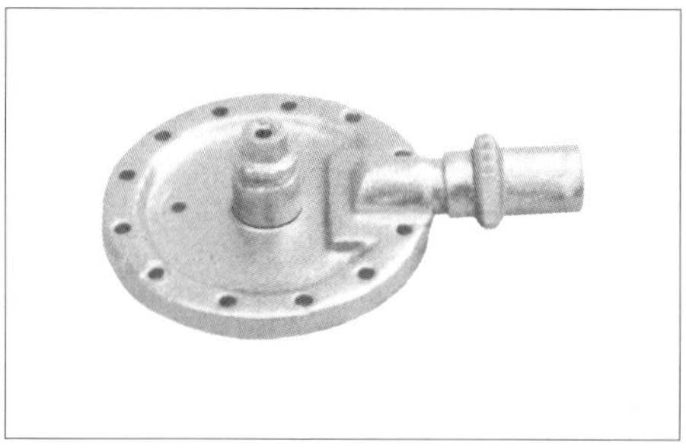

Photo 24: Displacer cylinder head with displacer rod gland, drilled for assembly

There are three other machining steps to be made: drilling the head from underneath to provide an air vent through to the outlet, drilling the holes (2.3mm) along the scribed line to take the 8 BA (or 2mm) bolts that secure the cylinder head to the cylinder flange and the last job is to drill for the a pressure escape vent on the opposite side of the outlet pipe – described in the section above dealing with the cylinder head. (Photo 24).

The pressure vent mentioned above is closed by a decorative screw, threaded 0 BA (6mm) and ending with a knurled top and fitted with a small 'O' ring as in Figure 7.20.

The next step in the assembly of the displacer unit is fitting the cylinder head to the displacer cylinder flange, marking and drilling the holes (using a 1.8mm drill) for tapping for the 8BA or 2mm bolts, keeping in mind that there should be a paper gasket, greased or lined with a thin layer of gasket cement, which also should be marked for the bolts.

18 DISPLACER, DISPLACER ROD AND CLEVIS

DISPLACER

(NOTE: The next important component is the displacer, a light weight air-tight chamber which shifts the volume of air from the hot side to the cold side. As far as can be ascertained the displacer in Stirling's model was made from a thin rolled iron sheet since it is reported that during a test it reacted to a magnet.)

The displacer is best machined from bright mild steel pipe 8½" = (216mm) long, O.D. 2 53/64+" = 2.834" (72mm), I.D. 2¾" = 2.756" (70mm). A drilled-through (hollow) wooden mandrel should be used and inserted tight fit into the pipe leaving a projection of 1" or 25mm when facing the pipe externally. With the mandrel still inside, the cylinder is sealed at one end with a brazed-on thin gauge mild steel disc. (The use of the mandrel is absolutely essential due to the thin wall resulting from both internal and external facing. (Figure 7.22)

Two methods can be used for sealing totally the displacer – a hollow light-weight aluminium plug or a brazed-on thin gauge mild steel disc which is first drilled 1.8mm to one side of the centre for safety, then drilled again centrally 2mm or 2.4mm to take the displacer rod. The first hole is tapped 8 BA and sealed with a short bolt with a smear of *Super Steel* adhesive after brazing. The centre hole is tapped 8 BA (alternative 2mm).

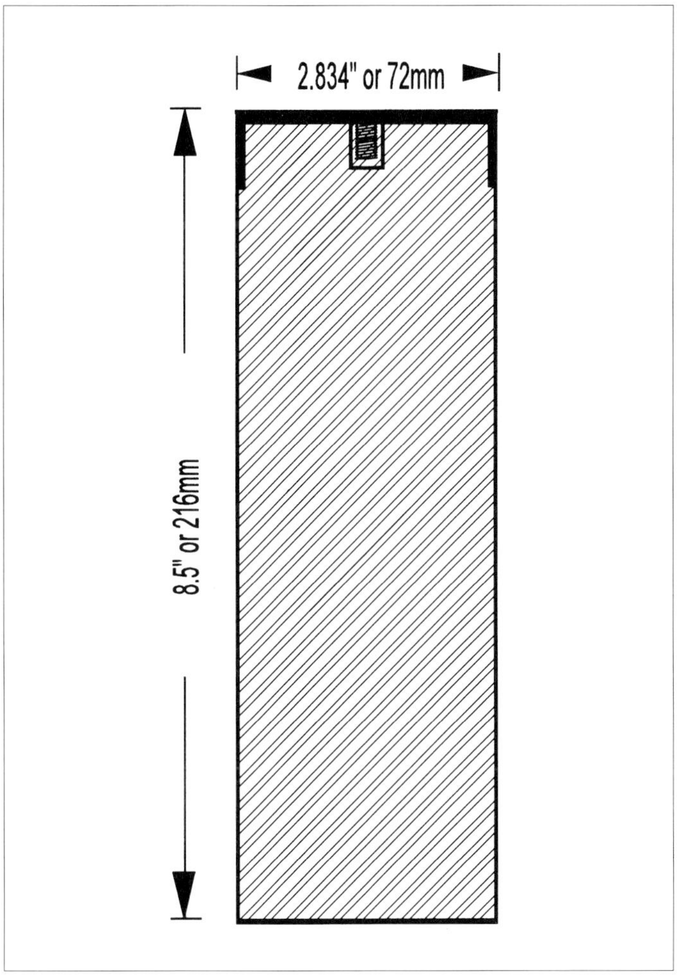

Fig. 7.22: Displacer

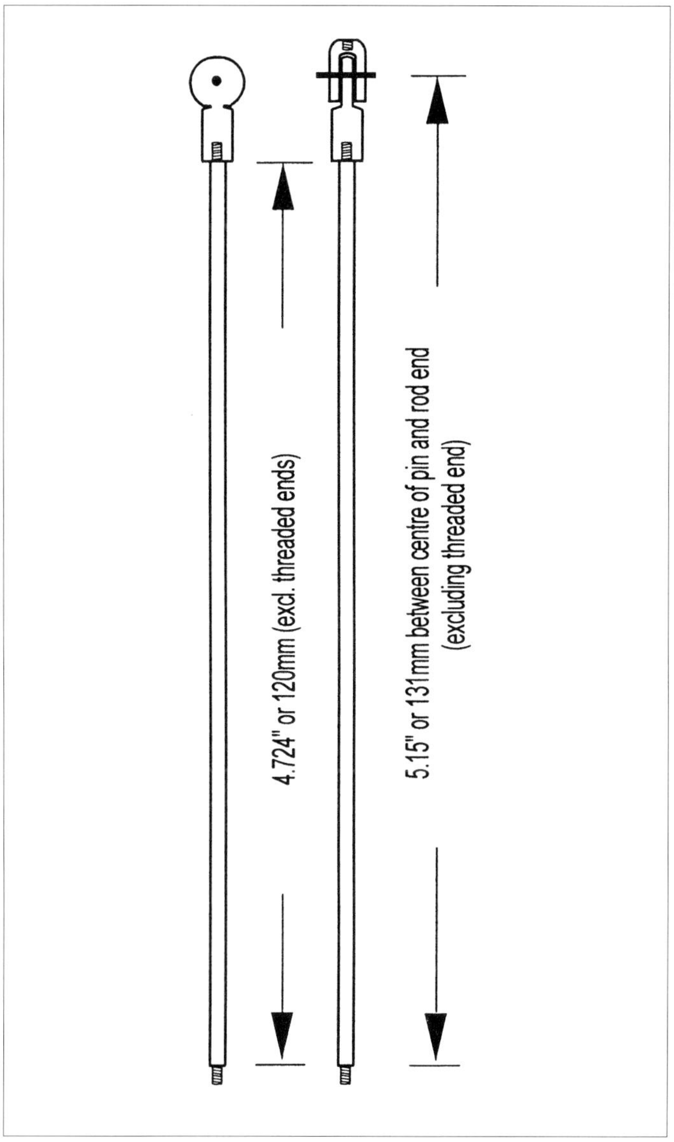

Fig. 7.23: Displacer rod (Actual length)

DISPLACER ROD

The displacer rod is cut from mild steel rod O.D. 1⁄8" = 0.125" (3.2mm) or 3.5mm , the length left at 5 1⁄32+" = 5.04" (128mm), reduced at both ends and threaded 8 BA (or 2mm) for a length of 5⁄32" = 0.156" or 4mm. At this stage the rod is temporarily fitted to the displacer until the actual length can be verified, while the other end is threaded to fit into the clevis bottom end, (Figure 7.23). Readers are reminded about the use of L/H and R/H taps and dies.

CLEVIS

NOTE: The wrist joint between the displacer con-rod and the displacer rod, a clevis-type device, is made from brass in two parts, the top part attached to the displacer con-rod, the bottom part to the displacer rod. The clevis in Stirling's model is somewhat decorative, not like the square or round types we are used to. Much of the work is bench work, with only a slot drilling and a threaded end. As the clevis is quite small it was decided to make enlarged (approx. 200%) drawings so that one could follow the finer details.

The clevis is cut from 6mm brass flat, filed down to 13⁄64" = 0.20" (5.12mm) thick, and shaped as in Figure 7.24. The round part which is 19⁄64" = 0.30" (7.64mm) in diameter ends in a 5⁄32" = 0.156" (4mm) neck then opens up diagonally to 9⁄32+" = 0.287" (7.3mm), which is the part that is threaded 8BA or 2mm. The clevis is first cross-drilled in the centre of the circle to take a 1.5mm pin, then slot-drilled 5⁄64" = 0.078" (2mm) up to the neck line.

The bottom part of this component that is screwed to the displacer rod, has a 1⁄4" = 0.25" (6.35mm) long neck, 5⁄64–" = 0.07" (1.8mm) thick that enters into the slotted end of the clevis. The bottom end, 5⁄32+" = 0.165" (4.2mm) long, is rounded end, drilled and tapped 8BA or 2mm. The total length of the bottom end of the clevis is 9⁄16" = 0.562" (14.3mm).

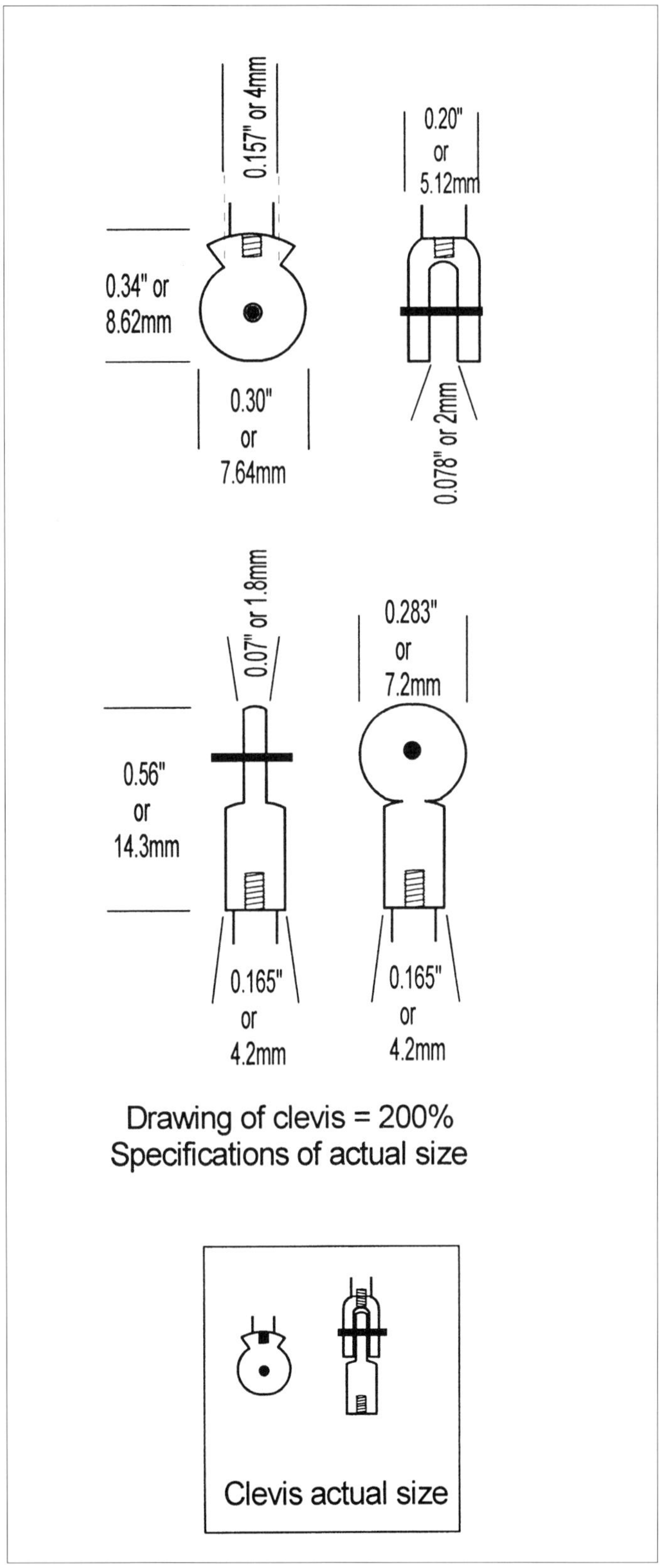

Fig. 7.24: Clevis-type device

NOTE: The following sections show Stirling's method of linking the crankshaft to the displacer; of particular interest is his arrangement of a sliding yoke to adjust the height of the displacer in its cylinder according to the stroke. Thus there was a uniform gap at the top and bottom of the stroke. A clear example is given in Keith Strong's Cad illustration of these components (Illustrations 1 & 2).

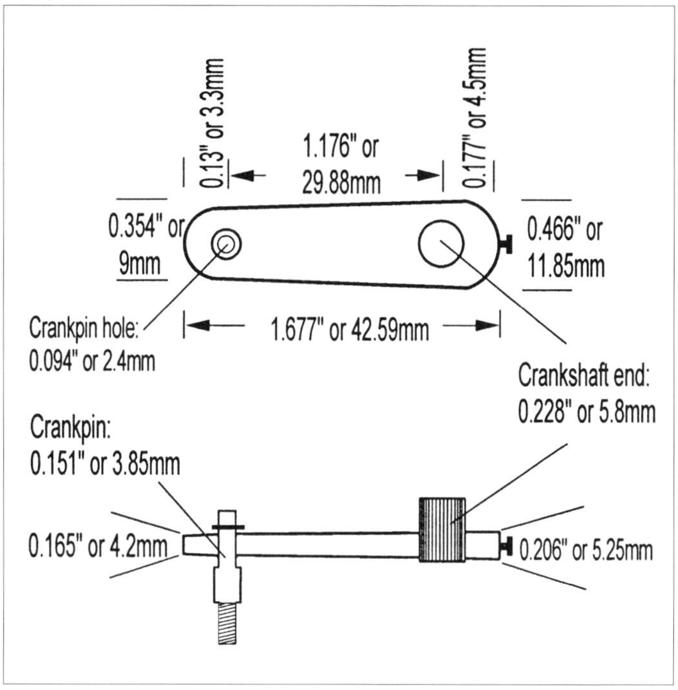

Fig. 7.25: Displacer crank (Actual size)

19 DISPLACER CRANK, CRANK PIN, YOKE AND CON-ROD

DISPLACER CRANK

The displacer crank is machined from brass flat 13/64+" = 0.206" (5.25mm) thick tapering to 5/32+" = 0.165" (4.2mm) in thickness, 1 43/64" = 1.677" (42.59mm) long, 15/32" = 0.466" (11.85mm) wide at the top reducing to 23/64–" = 0.354" (9mm) wide at the crank-pin end, rounded as in Figure 7.25.

The crank is marked at the top at a distance of 5/16" = 0.312" or 8mm from the top and marked again at a distance of 1 11/64+" = 1.176" (29.88mm). At the top mark the crank is drilled 7/32+" = 0.228" (5.8mm) to take the tapering crankshaft end, and cross-drilled and tapped 8 BA or 2mm to take a grub-screw to secure the crankshaft end. At the lower end of the crank, where marked, the crank is drilled 2.4mm to take the crankpin end tight fit.

(Actually a close up image of the crank shows that apart from the screw there is also a pin that goes through the crank and the crankshaft end.)

CRANKPIN

The crankpin is machined from bright mild steel rod. Reference is made to Figure 7.26 for the shape of the pin, part of which rotates in the crank, held by a pin, the central part secures the yoke while the remainder is threaded. The centre part has a shoulder against which rests a washer, the yoke, and then another washer. This centre part has an important role to play as it is the component that secures the yoke in position when the distance which the displacer moves inside the displacer cylinder has been carefully worked out and checked.

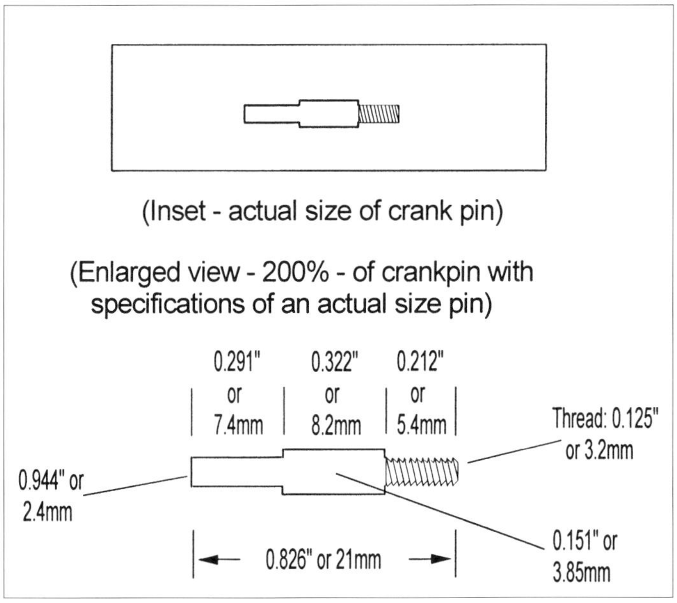

Fig. 7.26: Crank pin

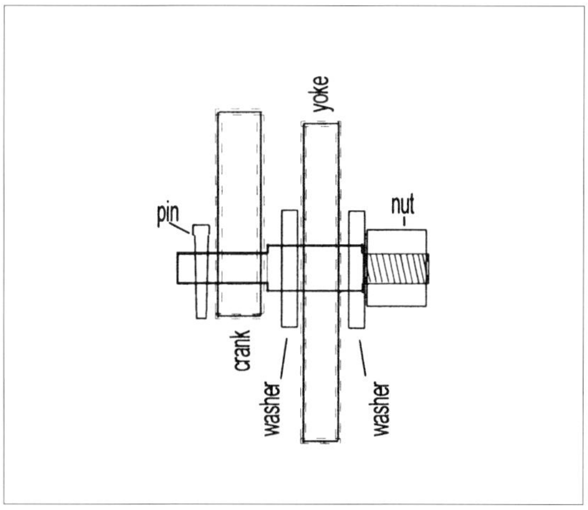

Fig. 7.27: Positions of components on crankpin (Slightly enlarged)

The overall length of the crankpin is 53⁄64" = 0.826" (21mm) long. A length of 9/32+" = 0.291" (7.4mm) is reduced 3⁄32+" = 0.0944" or 2.4mm, cross drilled at the end to take a 3⁄64" = 0.046" (1.2mm) pin. The distance where this hole is drilled depends on the actual thickness of the crank (plus the thickness of a thin washer which is advisable to prevent wobbling).

The centre part which is 21⁄64–" = 0.322" (8.2mm) long, has a shoulder 3⁄16" = 0.187" (4.8 – 5mm) in diameter. The thickness of the shoulder depends on the thickness of the two washers used and the thickness of the yoke. The remainder of the centre is reduced to 5⁄32–" = 0.151" (3.85mm).

The other end of the crankpin is threaded 1⁄8" = 0.125" or 3.2mm for a length of 7⁄32–" = 0.212" (5.4mm) to take a similarly threaded small square brass nut just 1⁄8" = 0.125" (3.2mm) long and 7⁄32" = 0.218" (5.5mm) square **.

Figure 7.27 shows the position of the crank and yoke on the crankpin.

*** In Stirling's model the nut and washer appear to be one whole part, machined in this manner.*

YOKE

The yoke is machined from 9⁄64–" = 0.137" (3.5mm) brass flat 25⁄64+" = 2.086" (53mm) long, 3⁄8" = 0.374" (9.5mm) wide. The piece is first reduced at one end for a length of 29⁄64" = 0.452" (11.5mm) and filed to a 1⁄8" or 3.2mm hexagon shape, drilled through and tapped 8 BA or 2mm to take the con-rod. The top part is marked and slot-drilled 5⁄32–" = 0.152" (3.88mm) for a length of 13⁄8–" = 1.372" (34.87mm). The yoke is then shaped as in Figure 7.28. *(If the machinist is using 3.5mm rods for con-rods, the end piece of the yoke should be left as the original thickness i.e. 9⁄64" or 3.5mm).*

The yoke (and the con-rod) adjusts its angle during the revolutions of the crank by means of the revolving crankpin.

DISPLACER CON-ROD

The total length (confirmed at source) of the con-rod, clevis and yoke in the original model is 4½+" = 4.508" (114.5mm) - Figure 7.29. Deducting the length of the yoke and the clevis, the actual net length of the con-rod is 25⁄32+" = 2.165" (55mm) including the two threaded ends, the top end threaded for a length of ¼" or 6mm and the bottom end threaded 3⁄32+" = 0.098" (2.5mm). The rod is cut from 1⁄8" (3.2mm) or 3.5mm mild steel and the ends threaded 8 BA or 2mm.

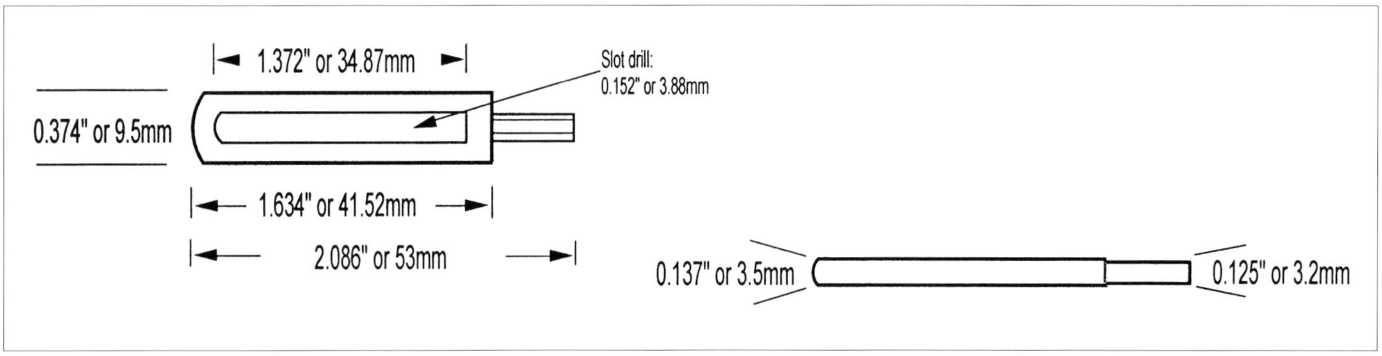

Fig.: 7.28: Displacer con-rod yoke (Actual size)

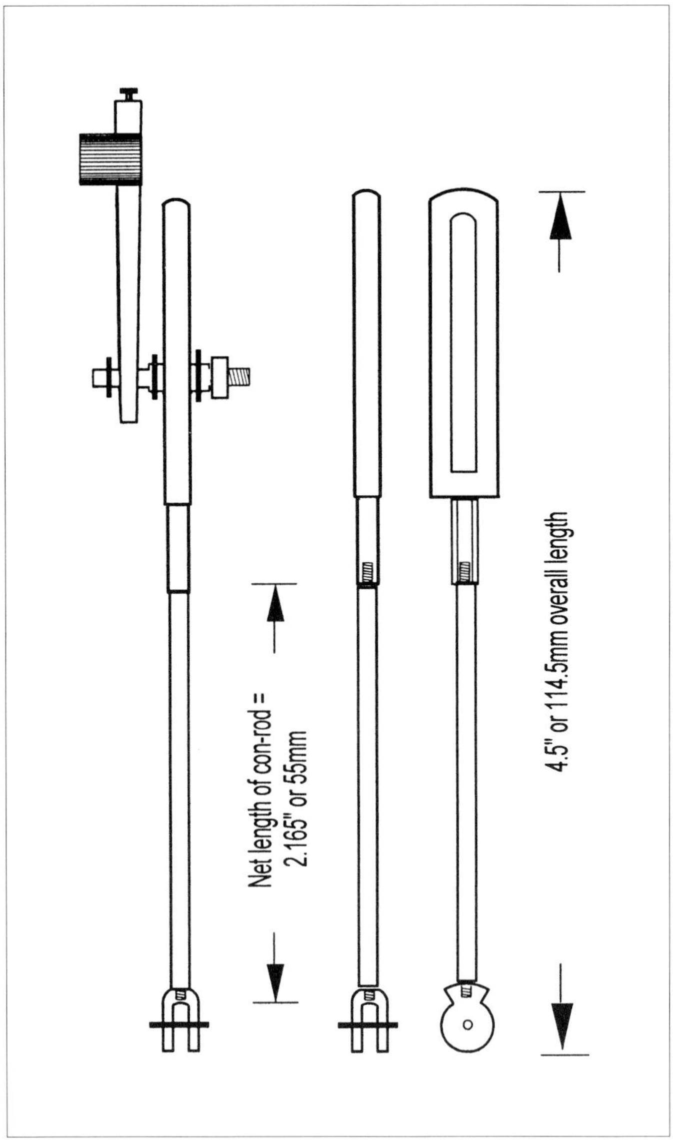

Fig. 7.29: Displacer crank, con-rod with yoke and clevis (Actual size)

20 FULL SCALE ASSEMBLY AND TESTING – STAGE 5

The displacer rod is screwed into the displacer with a smear of Super Steel adhesive and inserted into the displacer cylinder. The guide bush is screwed into the sealing flange with a thin "O" ring in between. The flange is bolted to the cylinder with a very thin gasket lightly smeared with grease, and finally the clevis bottom-end screwed on the displacer rod. The displacer unit is temporarily screwed to the engine base with the pressure outlet directly opposite and aligned with the pressure pipe from the power cylinder. (Figure 7.30 Photo 25).

The crankpin is inserted into the displacer crank which is in turn placed on the crankshaft end. The crank is set at 90° to the power crank, with the power crank at 12.00 and the displacer crank at 3.00 as a first trial.

The previously assembled displacer con-rod (yoke, rod and clevis top-end) is first attached to the displacer rod clevis with a 1.5mm pin. With the displacer crank at 6 o'clock, and the displacer in its lowest position (B.D.C.) in the cylinder, the yoke is mounted on the crankpin and lightly bolted (finger tight). The flywheel is given half a turn so that the displacer is now in its top position (T.D.C.). The crankpin bolt is loosened and the displacer pulled up manually to gauge the gap between its top position and the flange. The total gap should not be more than 2–3mm. The crankpin bolt is tightened again, this time reducing the gap to half its length, either 1 or 1½mm, so that the gap space is equal at the top and bottom of the stroke.

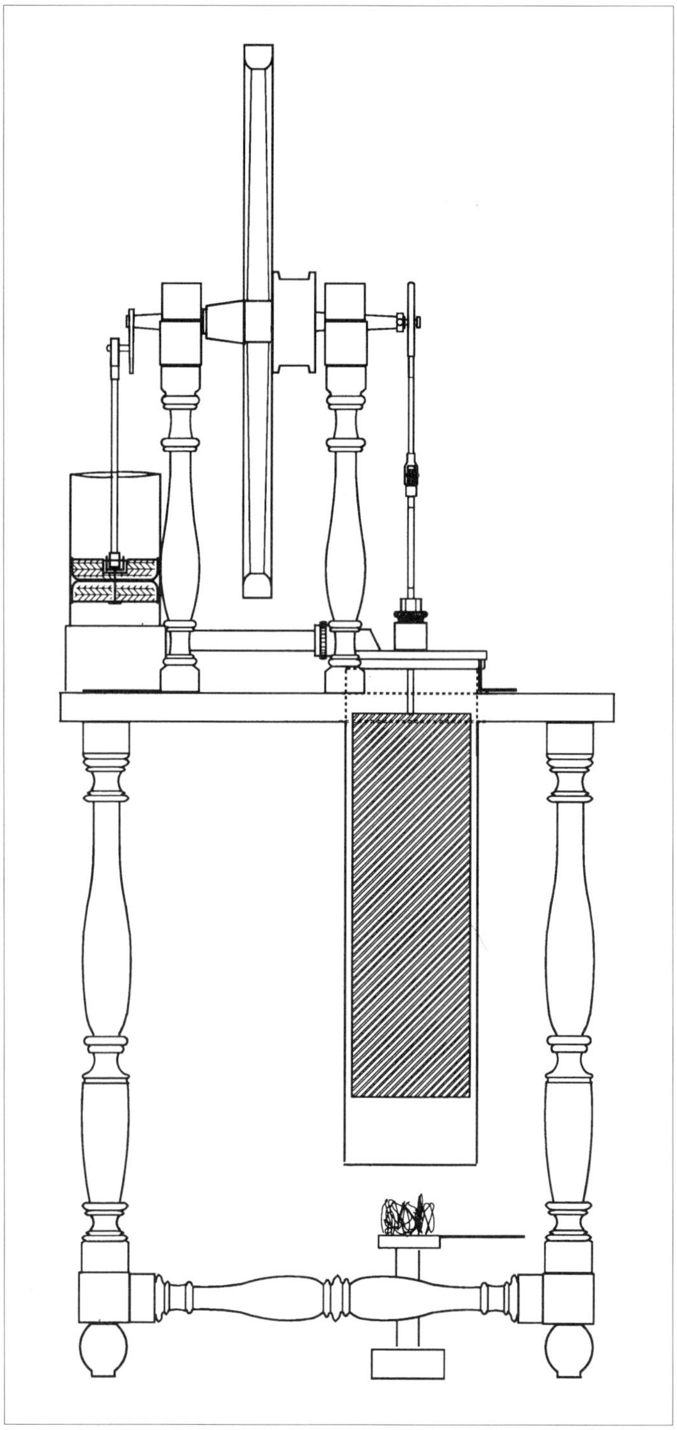

Fig. 7.30: Full assembly - Stage 5
Displacer unit on engine platform connected to power cylinder

The flywheel should be turned a few times with light lubrication in the moving/turning parts until satisfied that the components are working well.

All moving parts having been assembled and tested, it is time now to make a check-list of the model's components in the following order:

1 Flywheel assembly on lintels and the securing brackets screwed down fairly tight;

2 Power and displacer cranks assembled on crankshaft at 90°;

3 Displacer cylinder unit screwed into base; displacer con-rod into crankpin and bolted;

Photo 25: Full engine assembly of the replica Edinburgh model

Illustration No. 2: CAD drawing showing 4 stages of the crank/yoke system during a complete revolution of the flywheel (Courtesy - Keith Strong)

4 Power unit + base placed temporarily in position with the pressure pipe inserted into the connector from the displacer cylinder. Having made necessary alterations as to the length of pipe and including a smear of *Araldite* to seal the pressure pipe in place, the complete power unit is bolted to the engine base. (The model should be left lying on the displacer side overnight so that the Araldite will not run down on the base).

5 The power and displacer cranks are checked once more for the 90° phase.

BURNERS

Two alternative burners can be made for this engine, both of a simple spirit lamp-type device since very little heat is required to run the engine (keeping in mind that the displacer cylinder is made of brass and therefore heat spreads along its length very quickly). These kinds of common burners are relatively easy to make. Figure 7.31a.

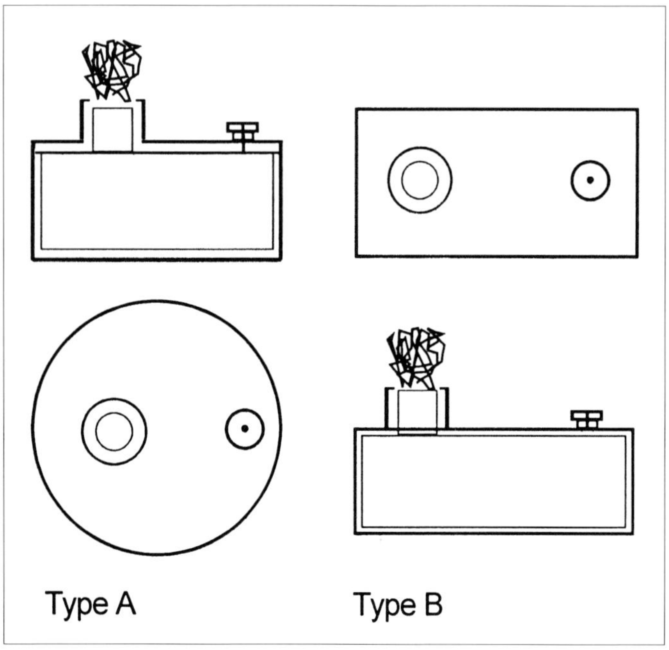

Fig. 7.31a: Types of spirit burners

Type A is bored and hollowed out from round brass 1⅜" = 1.378" (35mm) diameter, ⅝+" = 0.63" or 16mm thick, leaving an internal diameter of 1¼" = 1.25" or 32mm. A brass top is also machined from 35mm brass, ⅛" = 0.125" or 3mm thick, part of the thickness reduced to fit tight into the bored tank. It is advisable to keep some distance between the burner flame and the filler cap. For this reason the top is drilled off-centre in two places, on one side, ½" or 12mm from the edge, drilled either 8mm or 9mm; on the opposite side of the top, 7⁄32" = 0.218" (5.5mm) from the edge, an 0 BA or 6mm hole is drilled and tapped.

A spout is machined from ⅜" = 0.375" or 10mm brass rod, bored through 5⁄16" = 0.312" or 8mm. This can either be soldered to the top, or machined with a short step with a diameter of 9mm to fit into the top. A short filler screw, 0 BA or 6mm, is drilled 1⁄16" or 1.5mm though its length and inserted into the top. This is safety measure to allow heated spirit gas to escape from this hole. (Figure 7.31a), Photo 26.

Type B is a fabricated open (initially) rectangular brass container or tank, cut from 1⁄16" or 1.5mm brass sheet, folded into the shape shown in Figure 7.31b (without the top at this stage). The tank is silver soldered or brazed to ensure that no spirit is allowed to seep through the seams. A sheet from the same brass is drilled the same as the round burner, with the holes wide apart as shown, then brazed to the tank. It is advisable that the top sheet is cut slightly wider and longer than the tank, to ensure a good seal.

The spout and the filler screw are the same as in the round burner, with a slight alteration to the top. To ensure a good seat for the filler screw, an 0 BA or 6mm brass nut should be soldered to the tank surface (Photo 27). The filler screw should ideally have a fibre washer, as a rubber 'O' ring will not last long due to the heat generated by the gas flame, and conducted though the brass top.

Another safety device, (to avoid burnt fingers) is a handle to each of the above burners; this can be a short piece of 5⁄64" = 0.078" or 2mm brass, soldered to the side of either burner, with a wooden knob at the end.

Illustration No. 1: CAD drawing of the drive mechanism of the Edinburgh model showing the crank/yoke system. (Courtesy - Keith Strong)

Photo 26: Type 'A' spirit burner

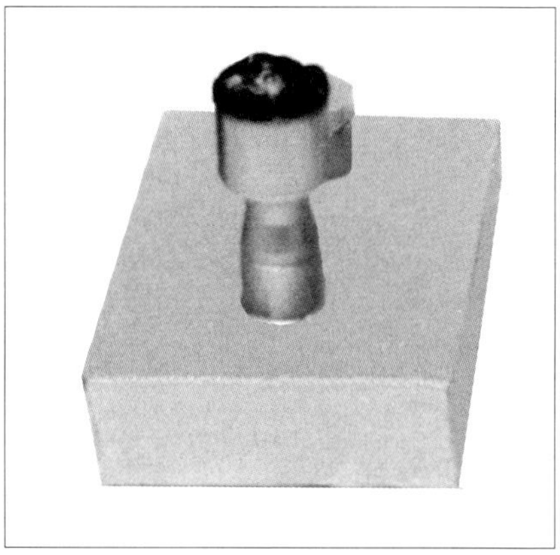

Photo 27: Type 'B' spirit burner

The cotton wick is inserted in the spout with enough length to make a coil in the bottom of the tank. If round wick the size of the spout is not readily available, a piece of flat wide wick is rolled tight into the spout, again with enough length to soak up the spirit.

Finally there is the minor problem of height between the bottom of the displacer cylinder and the floor level. If the burner is used solely for this model, then it is worth while re-designing the burner to include a stand as part of the equipment. In the case of Type A burner this could be fairly simple, a brass bar of 5/16" = 0.312" or 8mm threaded into the burner base (and sealed) and into a heavy round base, possibly the same diameter of the burner base. There should be at least 1" or 25mm separation between the top of the flame and the base of the displacer cylinder, (Figure 7.31b).

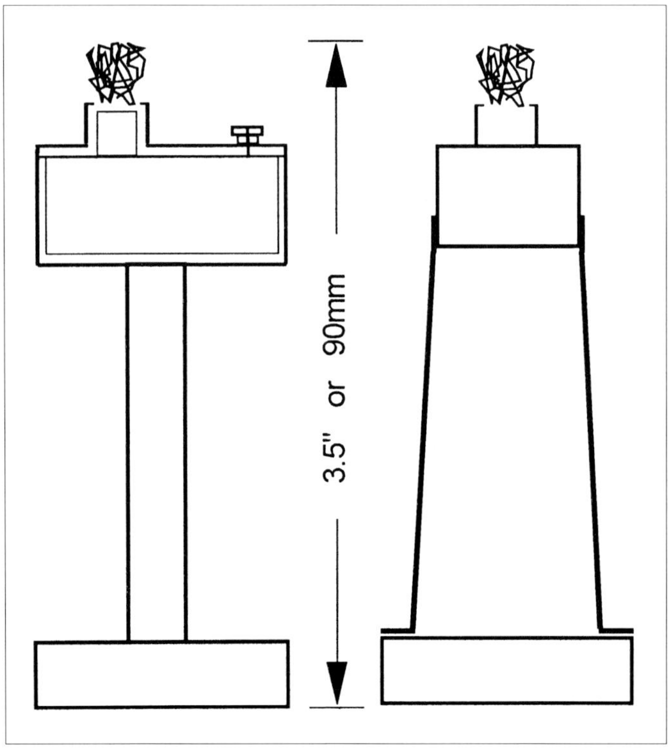

Fig. 7.31b: Stands adapted for burners

In the case of the rectangular burner, four legs, cut from brass flat, ½" or 12mm wide, 1/16" = 0 0625" or 1.5mm thick, slightly splayed outwards and soldered to the sides of the tank should suffice.

If the burners are likely to be used for other models the stands are designed differently, in which case a cradle or a tray, machined or fabricated from brass, and including legs as described above, will be sufficient.

Running the Edinburgh Model

The Edinburgh model's disadvantages in running for long duration stem from two different but related factors. In the first place the displacer cylinder (was) is made from brass, which heats up quickly and just as quickly travels along the length of the cylinder, with the obvious result that heat separation is lost to a large degree, the engine losses power quickly until it stops. If, on the other hand, the machinist elects to make the displacer cylinder from bright mild steel, the heat conduction is slower.

The other determining factor is that there is no cooling arrangement for the top part of the cylinder. All hot air/Stirling engines have this separation arrangement, the greater the difference between the hot and cold ends, the more efficient the engine becomes. In the case of the Edinburgh engine there is no physical separation and therefore the engine runs for a short while only. In fact care should be taken not to allow the displacer cylinder to become too hot at the top, and short runs only made with adequate cooling periods in between.

Robert Stirling's Model of the Air Engine at the Hunterian Museum, University of Glasgow

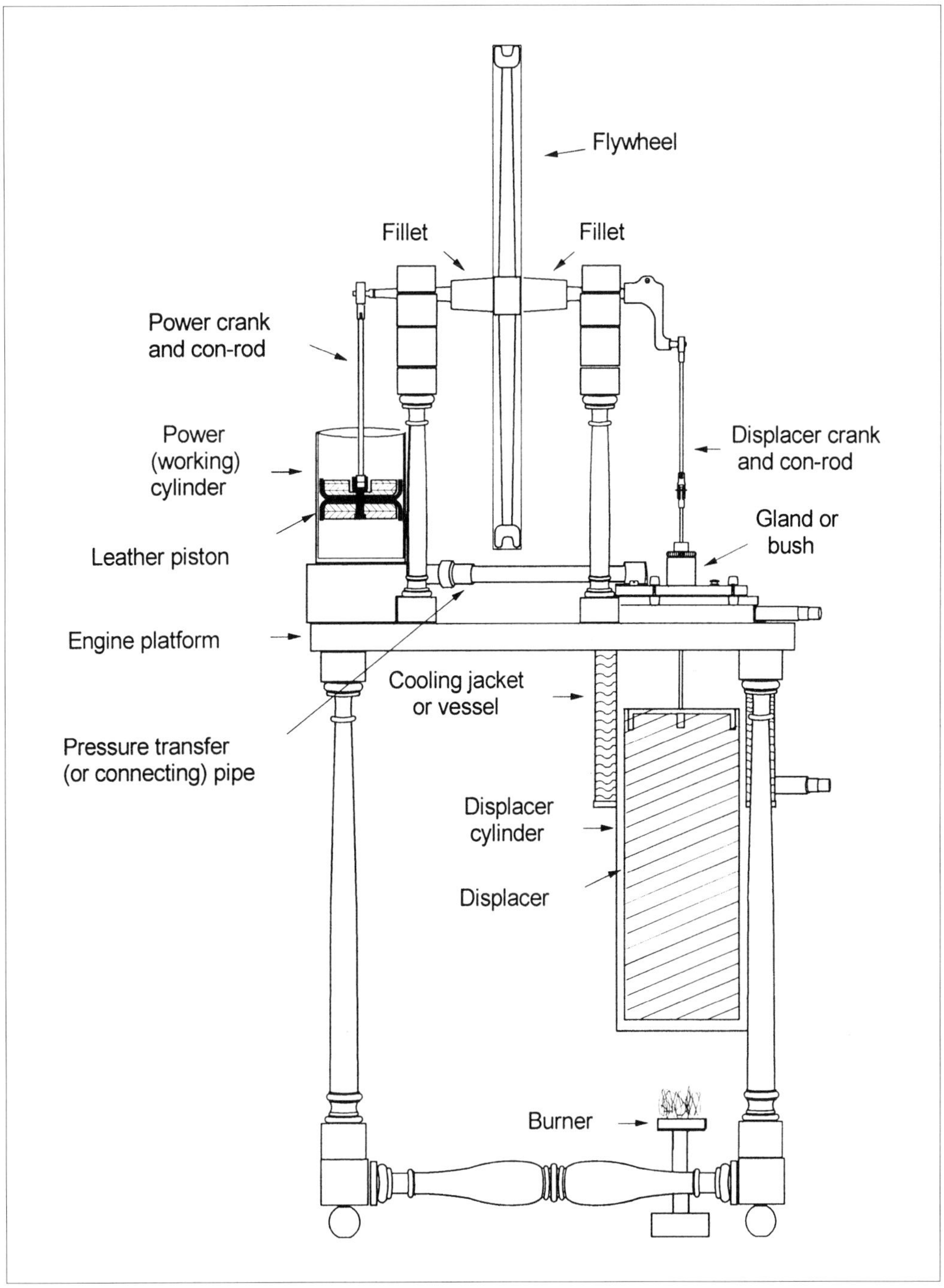

Profile of Robert Stirling's model of the 'Air Engine', held in the Hunterian Museum, University of Glasgow, Scotland

Chapter 8
Introduction - Glasgow Model

Although there are basic differences between the two models, the method of construction is in many ways similar and therefore easy to follow, machine and complete. The sequence also follows closely the one adopted for the Edinburgh model.

The wooden framework is practically the same in construction, four bottom columns on two cross-bars with ball-feet support the engine platform. Four top columns support a system of flywheel brackets on the lintels. The top and bottom columns differ in shape from the previous model in that these are more plain with less contours and therefore less complicated.

Starting from the top, the difference in the flywheel is that the rim of the Glasgow model is grooved to form a pulley, and instead of having a fillet and a pulley on the crankshaft, it has a fillet on either side of the flywheel.

Another difference is the connection between the flywheel displacer crank and the displacer con-rod – a more direct drive arrangement without any adjustment on a yoke.

The brackets securing the flywheel are different from the Edinburgh model, a three-part arrangement of lateral support.

The most obvious difference is the cooling water jacket around the displacer cylinder – it is hoped that one day someone will discover documentation whether this was part of the original engine as presented by Robert Stirling or an add-on when the engine was refurbished. This makes a difference in that either Robert Stirling or Professor Thomson realized the importance of effective difference in the temperature between the two ends of the displacer cylinder, in turn making the displacer a more effective regenerator and enabling the engine to run for longer periods.

One other difference is in the shape of the base cross-bars, square endings in this model however with the same type of ball-feet.

Some major differences are in the specifications and construction of the power and displacer units:

- the power cylinder of the Glasgow model has an O.D. of 2 5/32+" – 2.165" (55mm) and an I.D. of 2 3/64" – 2.047" (52mm) as against an O.D. of 2 1/16-" – 2.06" (52.32mm) and an I.D. of 1 61/64" – 1.95" (49.62mm) of the Edinburgh model;

- the power stroke is 1 37/64-" – 1.574" (40mm) as against 1 17/64" – 1.265" (32.15mm);

- the displacer cylinder has an O.D. of 3 7/32+" – 3.228" (82mm) and an I.D. of 3" – 2.99" (76mm) as against an O.D. of 2 53/64+" – 2.834" (72mm) and an I.D. of 2¾" – 2.756" (70mm);

- the length of the displacer cylinder is 10½" – 10.51" (267mm) as against 11¼" – 11.25" (285.75mm);

- the stroke of the displacer is 2 7/16+" – 2.44" (62mm) as against 2 23/64" – 2.35" (59.76mm);

- the length of the crankshaft is 7 51/64" – 7.795" (198mm) as against 6 19/64"– 6.3" (160mm)

- the ratio of displaced volume of gas (air) is 3.4:1 as against 3.5:1.

The construction of the displacer cylinders also varies – the Edinburgh model has a brass displacer cylinder and a rolled sheet iron displacer, while the Glasgow model has a rolled sheet iron cylinder and a rolled copper sheet displacer.

If one had to put an interpretation on the differences between the specifications and construction of the above components in the two models, one could surmise that in the short space of time between the presentation of the two models to the respective Universities, Robert Stirling, aided by his brother James, realized that certain deficiencies (albeit minor ones) in the first model had to be rectified to make the Glasgow model more efficient.

Historically it is known and recorded that when Professor Thomson found the model in the "Augean stables" it was in a poor condition and arranged for its rehabilitation. The University Library holds a large number of his laboratory account books, spanning over a number of years, and although these give the cost of repairs, they do not specify what the repairs were and therefore it has not been possible to identify which parts were repaired or replaced nor is it possible to speculate on what parts are the original and those that are not.

Information from the Hunterian Museum suggests there were many other alterations in the course of time and some of the minor components have been changed, soldered or replaced and therefore the profile of the original model may have altered slightly. Particularly obvious are the cranks at each end of the crankshaft which are completely different from any of the cranks in the previous end. Basically however the model still retains the characteristics of the master's genius.

Chapter 9
Woodwork - Glasgow Model

Construction Sequence 1 – 10

1 Engine platform,
2 Four top turned columns or pillars,
3 Lintels and flywheel securing brackets,
4 SUB-ASSEMBLY – stage 1,
5 Four lower columns,
6 Two balusters (horizontal turned bars),
7 Two base cross-bars and ball-feet,
8 SUB ASSEMBLY – stage 2,
9 Power cylinder platform
10 Fillet.

1 ENGINE PLATFORM

(NOTE: As in the case of the Edinburgh model the author suggests that a drawing of the engine platform, the position of the top columns. the bore, the power cylinder base, the power cylinder, together with an outline of the crankshaft is made on a sheet of paper or cardboard before marking the engine platform. In the case of the Glasgow model, which has been subjected to a great deal of shrinkage and warping the above mentioned drawing of the layout of each component is even more important. To make matters still more difficult for the modeller, there were quite a number of alterations due to repairs which must have made an effect of the original appearance. It has been difficult to measure the distance of the components from all angles and corners to arrive to a high degree of accuracy and in determining exactly the dimensions of the original model as constructed by Robert Stirling.)

The engine platform (Fig. 9.1a) is cut from mahogany: 45⁄64+" = 0.708" (18mm) thick, 11 31⁄32" = 11.97" (304mm) long and 6 9⁄64" = 6.14" (156mm) wide. A 4 41⁄64+" = 4.645" (118mm) hole is bored to take the displacer cylinder cooling jacket. The centre of this hole is 2¾+" = 2.75" (70mm) from the right hand edge. A slot ½" (12.7mm) wide is cut in the platform to allow for the cooling water jacket and the protruding pipes to slide into the platform.

The position of the top and bottom columns is shown in the plan view of the engine platform (Figure 9.1a and 9.1b) as well as on the front and side elevations (Fig. 9.2a and 9.2b). Figure 9.1c shows the bore in relation to the pedestals of the top columns and the capitals of the bottom columns, all four having to be chamfered to accommodate the water cooling jacket.

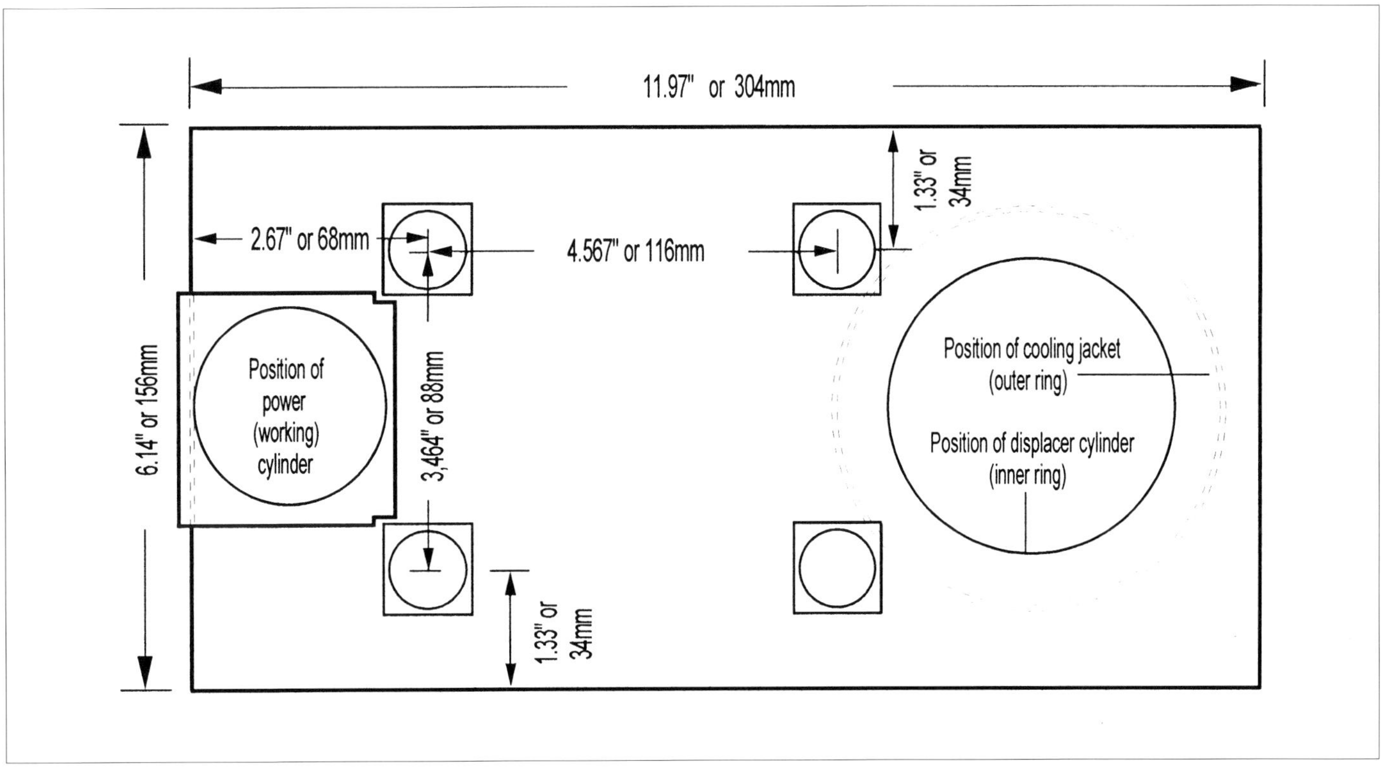

Fig. 9.1a: Engine platform (showing position of top columns, power cylinder platform, cooling jacket and displacer cylinder)

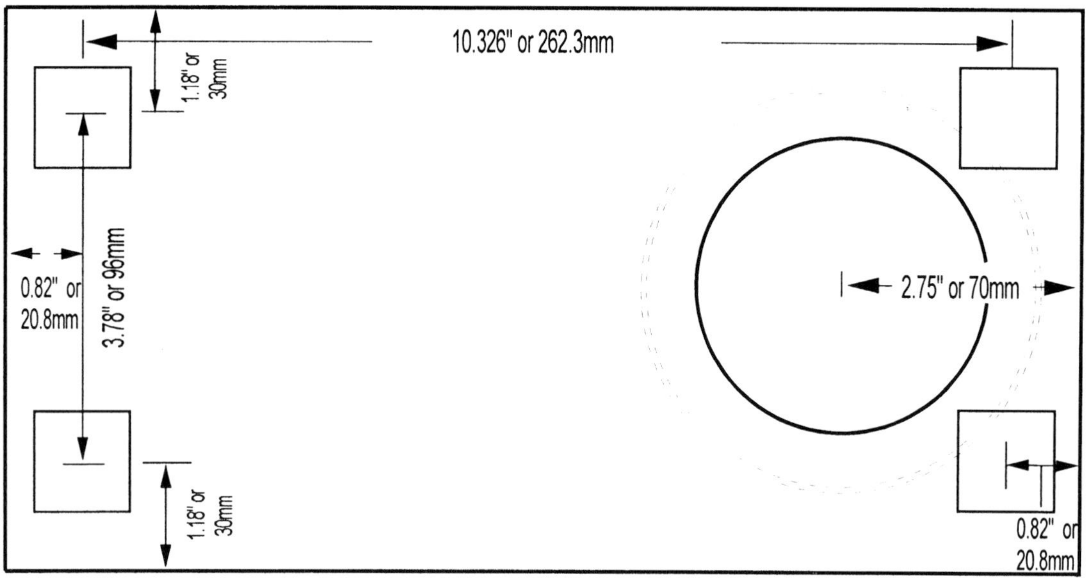

Fig. 9.1b: Engine platform (showing position of bottom columns, also in relation to cooling jacket)

The centre of the top LEFT hand columns is 2⁴³⁄₆₄" = 2.67" or 68mm from the L/H edge, and 1²¹⁄₆₄" = 1.33" (OM 34mm) from the front and rear edges. The centre of the LEFT hand bottom columns is ⁵³⁄₆₄" = 0.82" or 21mm from the L/H edge and 1³⁄₁₆–" = 1.18" or 30mm from the front and rear edges.

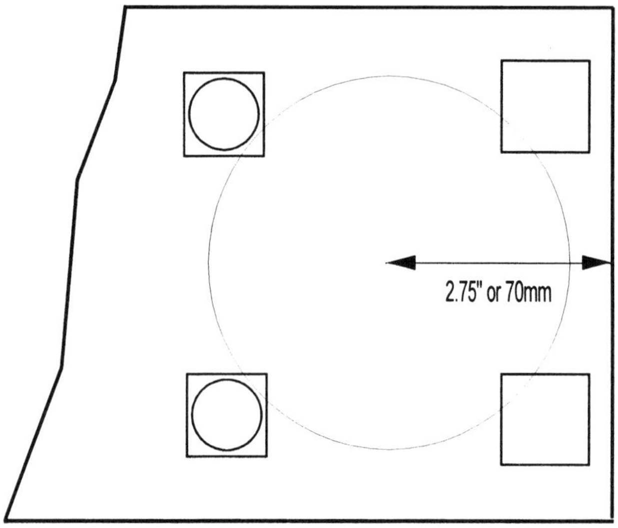

Fig. 9.1c: Cooling jacket bore in relation to the pedestals and capitals of the top and bottom columns

The platform top is drilled ³⁄₁₆" or 5mm and 5mm deep in the centre of the four squares (circled inside – representing the pedestals) to take dowels for the top columns; the underside is similarly drilled in the centre of the larger squares (representing the capitals of the bottom columns).

(NOTE: The bore for the cooling jacket is very close to the drilled holes which take the right hand pair of columns, and therefore care should be taken both in drilling these particular holes and well as in fitting the columns which need special attention – see description on columns.)

In each case the holes are drilled to take ³⁄₁₆" or 5mm dowels as are the top and bottom columns. The platform is sanded, given a coating of mahogany stain, and then clear semi-matt varnish.

(All other wood pieces were treated in this manner to give a uniform colour.)

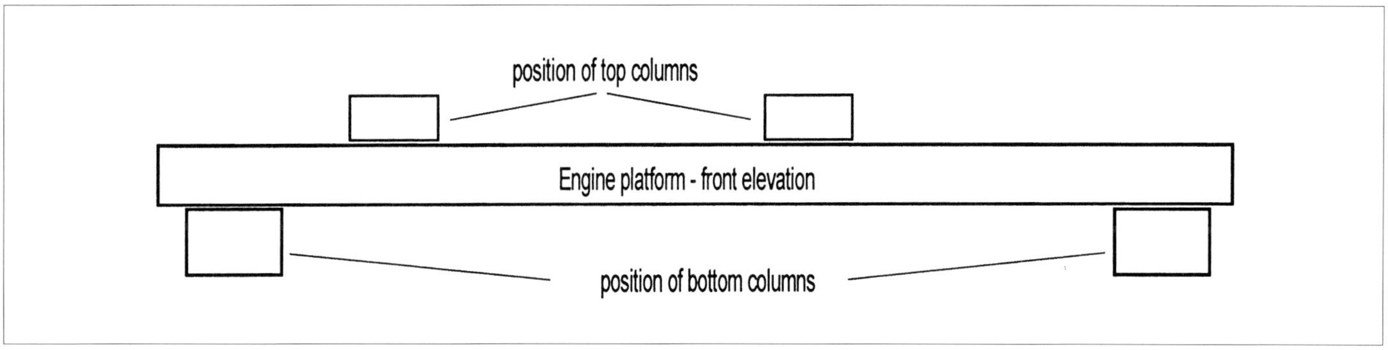

Fig. 9.2a: Front elevation of engine platform

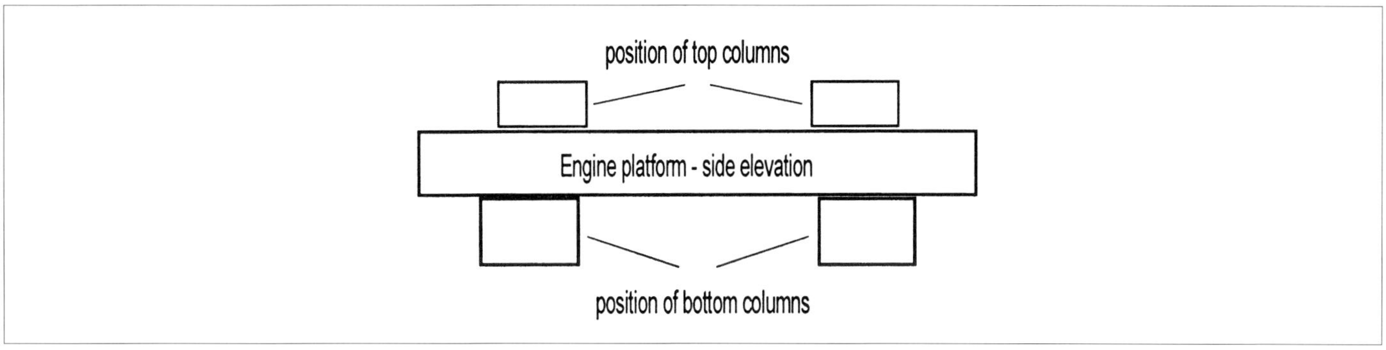

Fig. 9.2b: Side elevation of engine platform

2 TOP TURNED COLUMNS

The top four turned columns are $6\frac{3}{32}$" = 6.10" (155mm) long; the capitals and pedestals are $\frac{63}{64}$" = 0.984" (25mm) square, and $\frac{19}{32}$" = 0.59" (15mm) high. These are machined from square pieces of mahogany or beech, first marked 15mm from each end, leaving $4\frac{59}{64}$" = 4.92" (125mm) to be shaped as in Figure 9.3, (Photo 28).

The columns are centre-drilled at the top and bottom to take $\frac{3}{16}$" or 5mm dowels to a depth of $\frac{3}{16}$" for fitting into the lintels and the engine platform. The two right hand columns have their pedestals chamfered in the corners close to the bore to allow for the water cooling jacket to slide into the engine platform. The amount of chamfer or cut into the pedestals should be measured during a trial assembly of these columns.

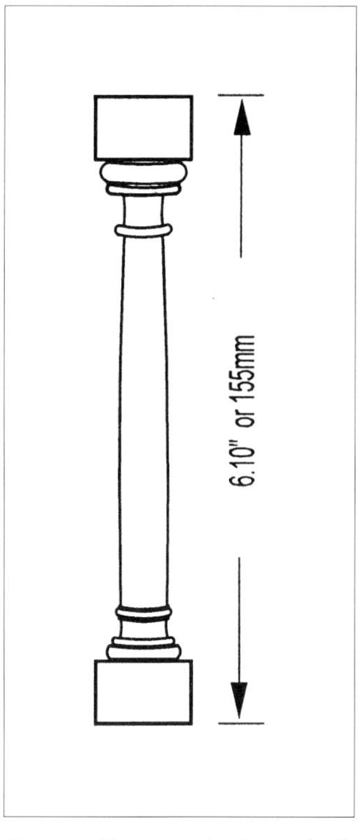

Fig. 9.3: Top turned column (x 4)

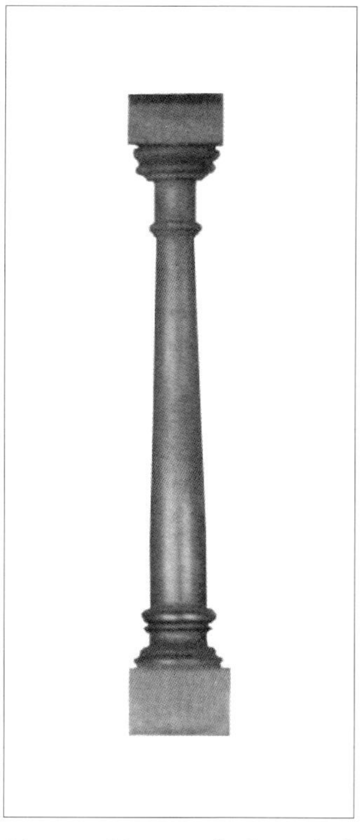

Photo 28: Top turned column (x 4)

3 FLYWHEEL SUPPORTING BRACKETS

NOTE: In the Glasgow model Robert Stirling opted for a different system of securing the flywheel crankshaft. In the Edinburgh model this rests between two sets of wooden brackets, each set consisting of a lintel which sits on two top turned columns, and a securing bracket which holds the crankshaft rigidly in place. In the Glasgow model the system consists of a lintel and two bearing brackets, a bearing base and a bearing cap with a set of bearing shells (either copper or lead) in between. The lintel is separate from the bearing blocks and is apparently glued to the top columns; however each pair of bearing brackets are bolted to the lintel.

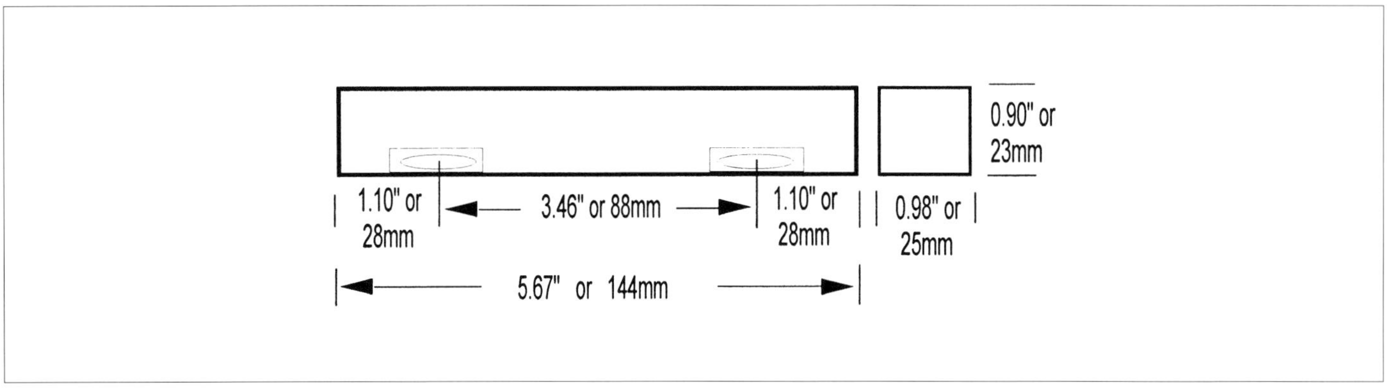

Fig. 9.4a: Lintel

The LINTELS are cut, planed and finished from hardwood pieces $5\frac{43}{64}$" = 5.67" (144mm) long, $\frac{63}{64}$" = 0.984" (25mm) wide, $\frac{29}{32}$" = 0.90" (23mm) high, then marked underneath to take the top columns, each centre being $1\frac{3}{32}$+" = 1.10" (28mm) from each end, and $3\frac{15}{32}$" = 3.46" (88mm) between the two marks. (Fig. 9.4a)

The BEARING BASES are cut, planed and shaped from hardwood pieces $2\frac{9}{16}$" = 2.56" (65mm) long, $\frac{63}{64}$" = 0.984" (25mm) wide and $\frac{3}{4}$–" = 0.748" (19mm) high, (Fig.9.4b). At this stage no more work is done until the bearing caps are machined.

The BEARING CAPS are also cut, planed and shaped from hardwood, $2\frac{23}{64}$" = 2.36" (60mm) long, $\frac{15}{16}$+" = 0.94" (24mm) wide and $\frac{5}{8}$+" = 0.63" (16mm) high. The bearing caps in Stirling's model were given a distinctive shape, each end being reduced in width and rounded bow-shaped as in Fig. 9.4c.

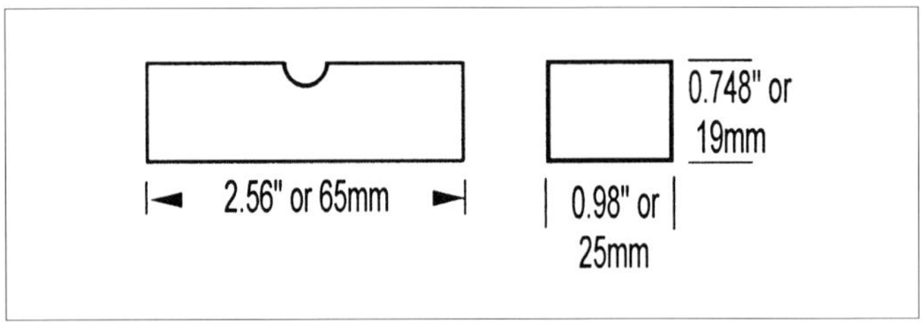

Fig. 9.4b: Bearing base

The lintels, bearing bases and caps are then drilled together to take 4mm countersunk screws, $1\frac{27}{64}$–" = 1.417" (36mm) between centres and equidistant from each end.

The bearing caps and bases are first screwed together, then cross-drilled in the centre to take the crankshaft end. At this stage the hole is drilled $\frac{7}{32}$" or 5.5mm. Further work is required on these two sets of brackets, once the bearing shells are constructed (see Chapter 11, sequence 12). The bearing brackets and the lintels are then temporarily bolted together (Fig 9.5).

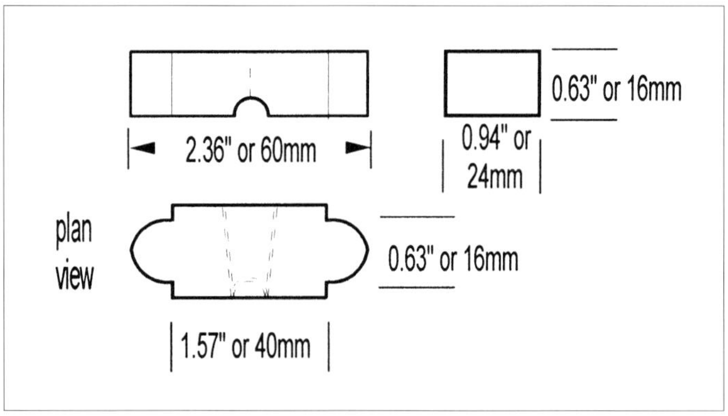

Fig. 9.4c: Bearing cap

4 SUB-ASSEMBLY – Stage 1

As with the Edinburgh model a trial assembly is made with the top turned columns first with dowels into the engine platform, then into the lintels with the bearing brackets in place and a rod replacing the crankshaft. Once it is ascertained that the components fit in place and are square with the platform, the columns are glued with *Araldite* or wood glue into the platform only (Fig. 9.6) (Photo 29).

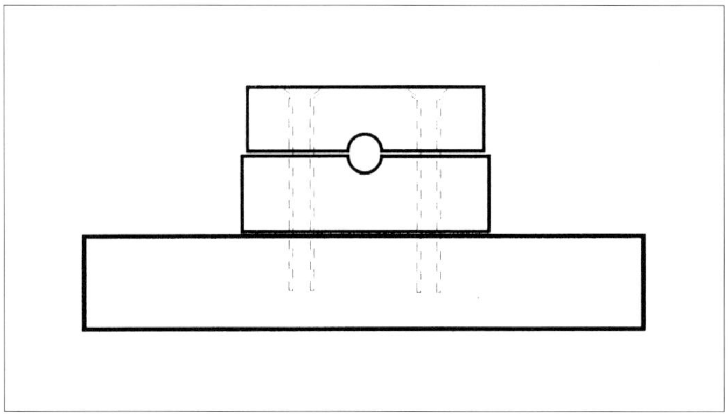

Fig. 9.5: Crankshaft supporting brackets

5 BOTTOM TURNED COLUMNS

The four bottom turned columns that support the engine platform are $12\frac{1}{8}$" = 12.125" (308mm) long; the capitals are $1\frac{3}{32}$+" = 1.10" (28mm) square and $\frac{3}{4}$" = 0.748" (19mm) high, while the pedestals are also 28mm square and $\frac{53}{64}$" = 0.826" (21mm) high.

The columns are machined from square pieces of mahogany or beech, first marked 19mm from the top end, and 21mm from the bottom end leaving $10\frac{35}{64}$+" = 10.55" (268mm) to be shaped as in Figure 9.7, (Photo 30). The capitals of the right hand columns are also chamfered in the corners close to the bore in the engine platform – this to allow the insertion of the water cooling jacket.

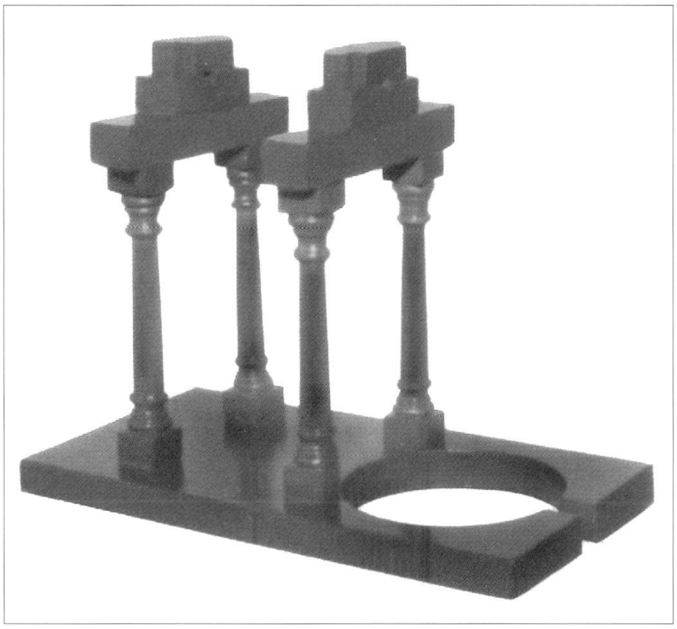

Photo 29: Sub-assembly of top wooden structure

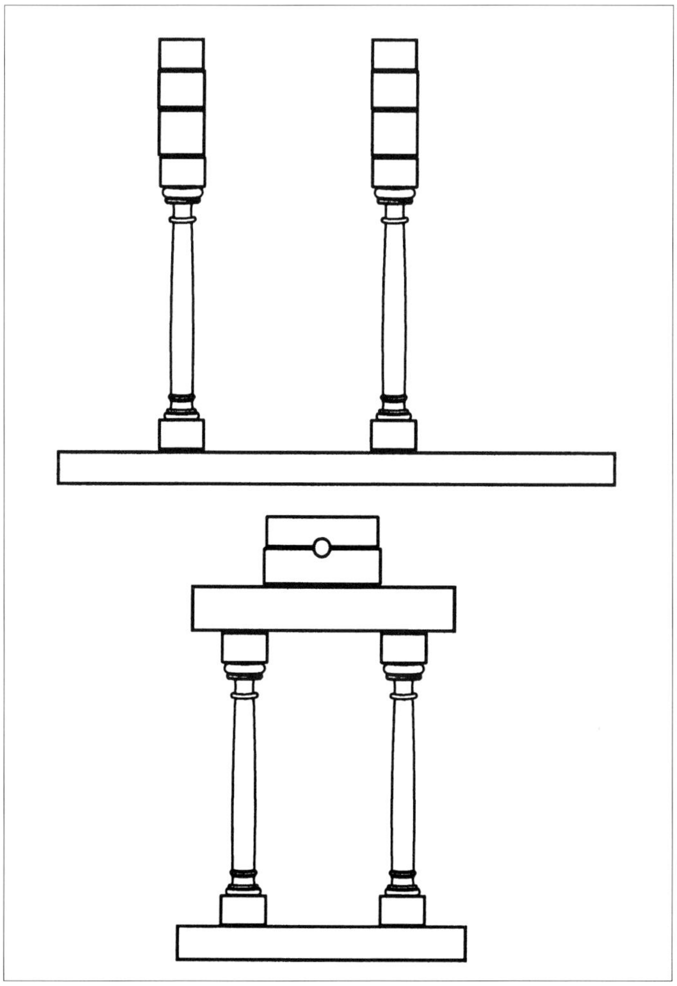

Fig. 9.6: Sub-assembly - Stage 1 Front and side elevation of top structure

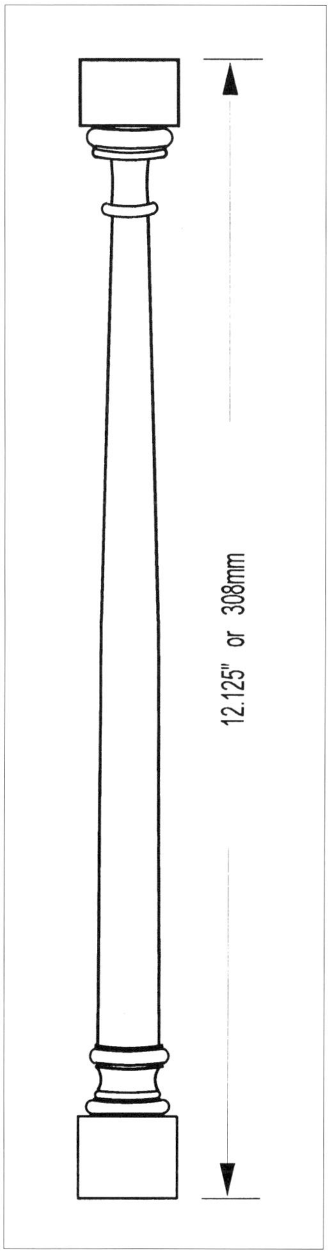

Fig. 9.7:
Bottom turned column (x 4)

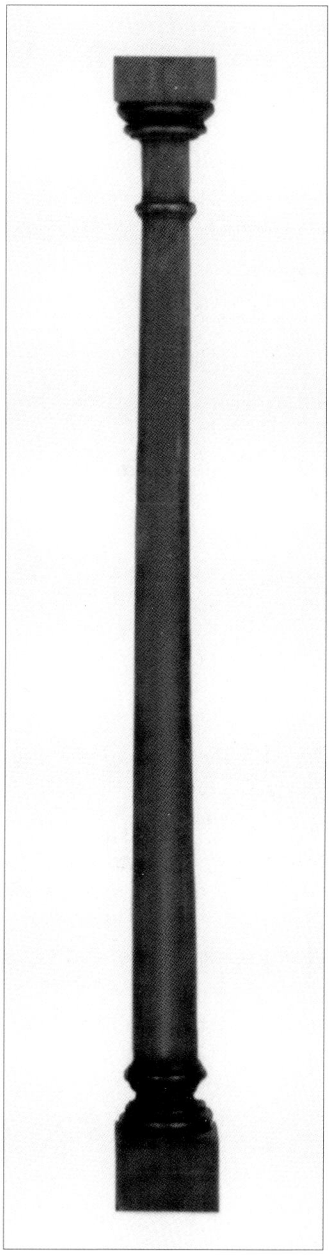

Photo 30:
Bottom turned column (x 4)

The columns are centre-drilled at the top and bottom to take $\frac{3}{16}$" or 5mm dowels to a depth of $\frac{3}{16}$" for fitting into the base cross-bars and the engine platform.

6 BALUSTERS (HORIZONTAL TURNED BARS)

(NOTE: The balusters in the Glasgow model are slightly different in the way the ends finish. Whereas the Edinburgh model has square ends – like capital/pedestals, the Glasgow model end in a round contoured shape; however there is a narrow square fillet or packing piece at each end. It is difficult to surmise why there should be packing pieces, unless of course the balusters were machined too short and therefore these made up for the difference in length. The balusters for this replica are therefore machined to the same length of the original and similar packing pieces put in place at each end.)

The balusters are $8\frac{55}{64}$" = 8.86" (225mm) long with a maximum diameter of $1\frac{1}{16}$" = 1.063" (27mm) in the contours, shaped as in Figure 9.8. Each end is centre-drilled and drilled $\frac{3}{16}$" or 5mm to take a dowel. (Photo 31).

The packing pieces are cut from thin wood (plywood) $\frac{1}{8}$+" = 0.137" (3.5mm) thick, $1\frac{1}{16}$" = 1.063" (27mm) square, drilled in the middle to take the dowels. The total length of the balusters and packing pieces is $9\frac{1}{8}$+" = 9.133" (232mm)

7 BASE CROSS-BARS

The two base cross-bars that form the base of the whole model are cut $11\frac{7}{32}$" = 11.22" (285mm) long, $1\frac{3}{32}$" = 1.10" (28mm) wide and $1\frac{1}{16}$"* = 1.063" (27mm) high, square-end shaped as in Figure 9.9. They are marked on the top surface $3\frac{3}{4}$–" = 3.74" (95mm) from each end of the surface, and 95mm between centres; centre-drilled $\frac{3}{16}$" or 5mm for a depth of $\frac{3}{16}$" to take similar size dowels. These take the four bottom columns.

The cross-bars are turned over, the bottom marked $1\frac{3}{16}$–" = 1.18" (30mm) from each end and drilled $\frac{3}{16}$" to take similar size dowels to a depth of $\frac{3}{16}$". These take the four semi-rounded ball-feet, O.D. $\frac{3}{4}$" = 0.75" (19mm) and 19mm high, on which the cross-bars rest. The ball-feet are similarly drilled to take the dowels. At this stage the ball-feet are glued and inserted into the base cross-bars.

The cross-bars are then turned on their side so that the inner side of the left bar is facing the inner side of the right. Both are marked $2\frac{13}{64}$" = 2.2" (56mm) from each end and drilled to take similar dowels for the balusters.

8 SUB-ASSEMBLY – STAGE 2

The balusters are fitted to the cross-bars, followed by the bottom turned columns which in turn are then fitted into the engine platform (Figure 9.10). The top and bottom halves of the engine structure are then glued together as in Figs. 9.11a and 9.11b (Photo 32).

9 POWER CYLINDER PLATFORM OR BASE

NOTE: The power cylinder base or platform is not considered to be the original. In the Edinburgh model, the complete power unit (cylinder and base) fit within the left hand top columns. The Glasgow 'replacement' base was cut wider that the distance between the columns, and had therefore to be adapted by removing parts from the front end in order to fit in between the columns. Even so the platform still projects by about $\frac{1}{8}$" or 3mm over the engine platform, giving it a rather untidy look. The machinist may consider reverting to the original and copying the Edinburgh model's power cylinder platform but with dimensions that fit within the columns and on the engine platform correctly.

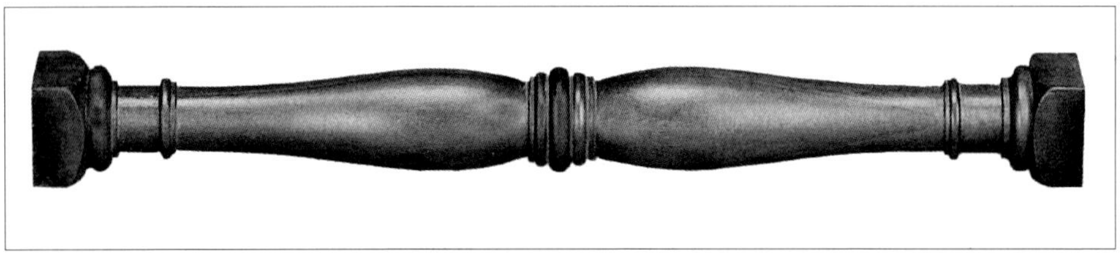

Photo 31: Baluster (x 2)

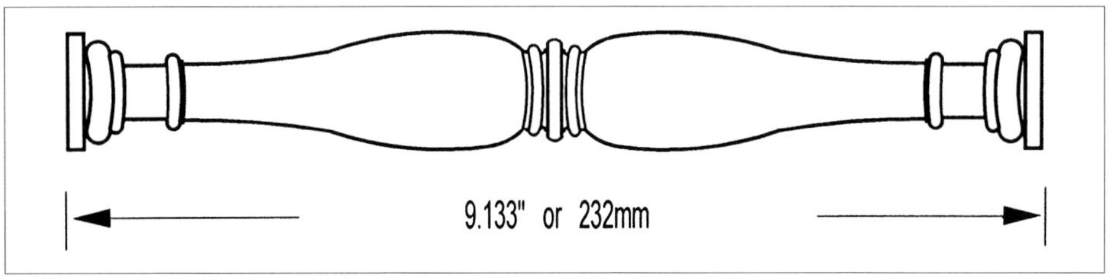

Fig. 9.8: Baluster (with packing pieces added on at each end)

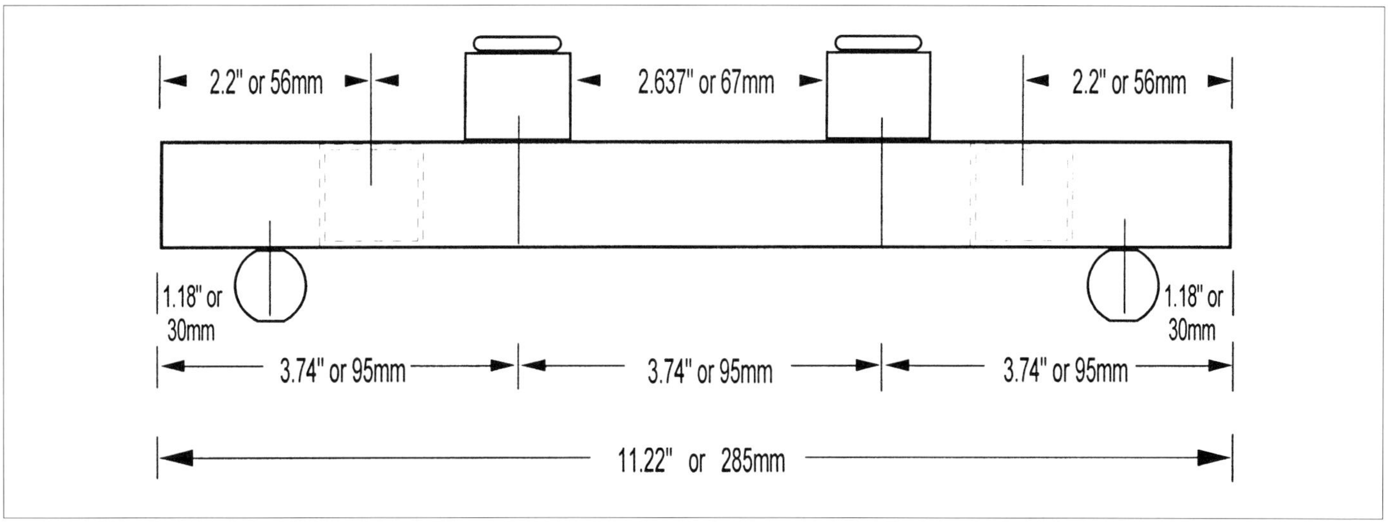

Fig. 9.9: Base cross-bars with positions of columns, ball-feet and balusters

According to information given to the author, the power cylinder (with the base) has a sliding adjustment which allows (or rather allowed) the cylinder to be removed by loosening a screw in a slot in the engine platform and sliding the cylinder off the end of the connecting pipe. Apparently a socket/sleeve extension was added to the pipe from the power cylinder when the engine was being rehabilitated.

Therefore there remains the question of whether to introduce the sliding arrangement through the slot in the engine platform. This is not considered necessary because apart from the fact that these were obvious measures to repair the damage done during the 20 years of neglect, careful calculations of all the parts between the two cylinders will eliminate these corrective measures and restore the replica to the original state which Robert Stirling donated it to the University of Glasgow.

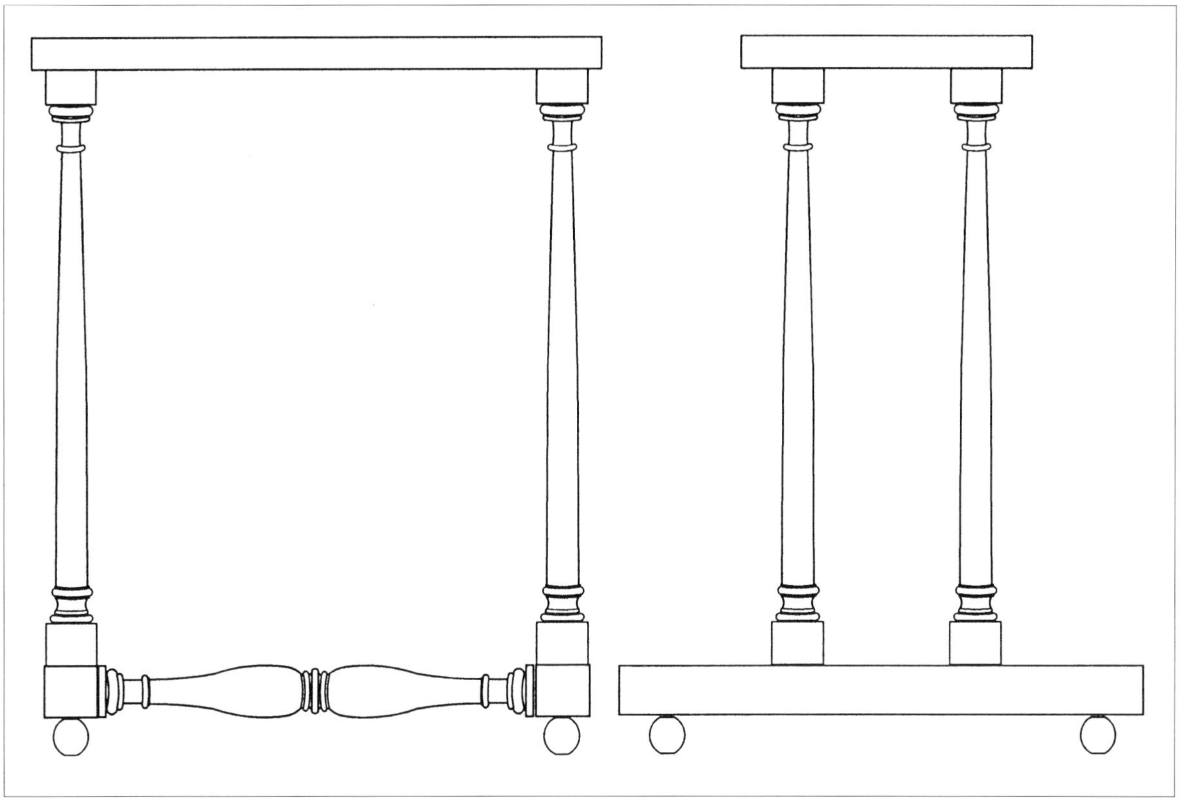

Fig. 9.10: Sub-assembly - Stage 2 Assembled bottom structure (1/4 scale)

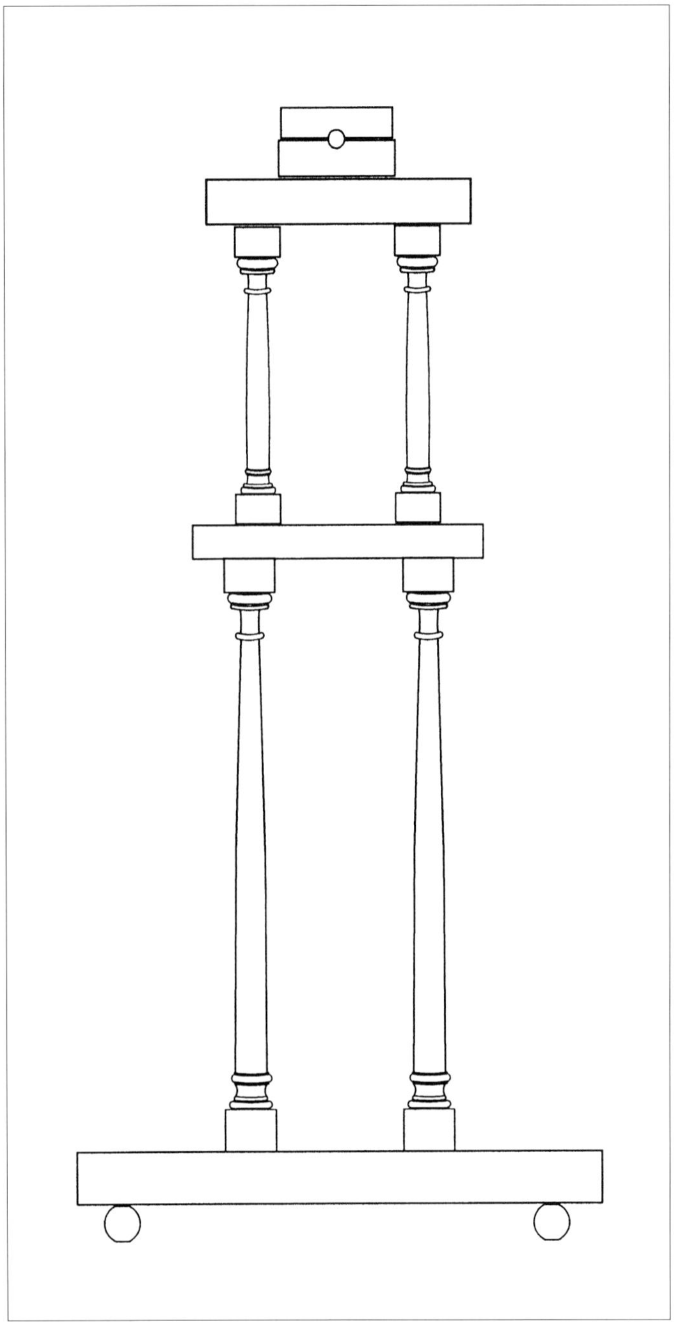

Fig. 9.11a: Side elevation of assembled wooden structure (¼ scale approx)

Fig. 9.11b: Front elevation of assembled wooden structure (¼ scale approx)

The wooden platform or base beneath the power cylinder is cut and finished from mahogany wood with the following dimensions: 2³¹⁄₆₄" = 2.48" (63mm) wide (as seen from the front), 2⁹⁄₁₆" = 2.56" (65mm) long (as seen from the side) and 1¹³⁄₃₂+" = 1.417" (36mm) high. A piece 3mm × 2.5mm was cut from both sides of the front end – the isometric image and the side elevation show how the replacement cylinder base appears (Fig. 9.12).

The surface of the block is hollowed out to accommodate the pressure transfer pipe from the displacer cylinder which goes into the base of the power cylinder. The part that is hollowed is only 1" (25mm) long and the diameter of the pipe. Finally the platform is drilled centrally to take a 4mm countersunk bolt.

10 FILLETS

The Glasgow model has two fillets one on each side of the flywheel hub. These are machined from mahogany wood ¹³⁄₁₆"= 0.811" (20.6mm) diameter, 1¹⁄₆₄" = 1.063" (27mm) long, tapering to 45/64+" = 0.708" (18mm) and shaped as in Figure 9.13. The two fillets together with the flywheel measure 2¾+" = 2.756" (70mm).

***Note: see description of crankshaft and drill the fillets accordingly.*

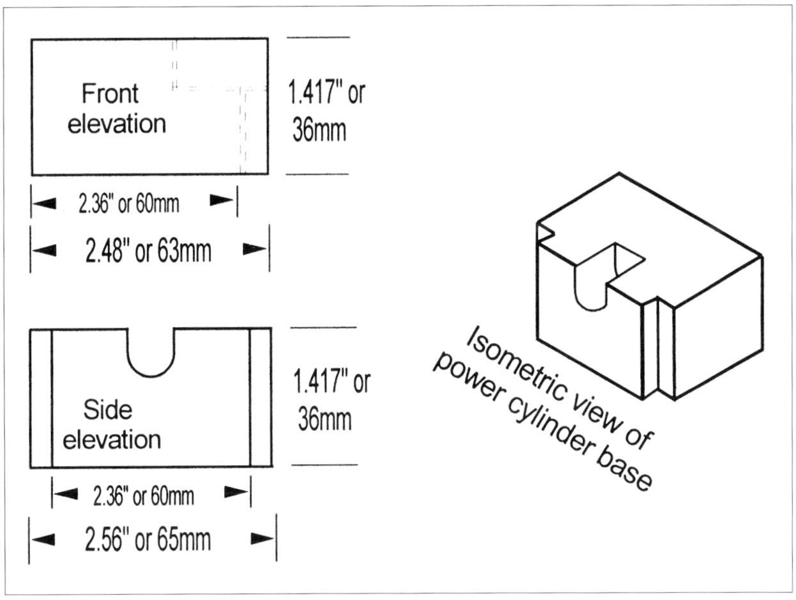

Fig. 9.12: Power cylinder platform

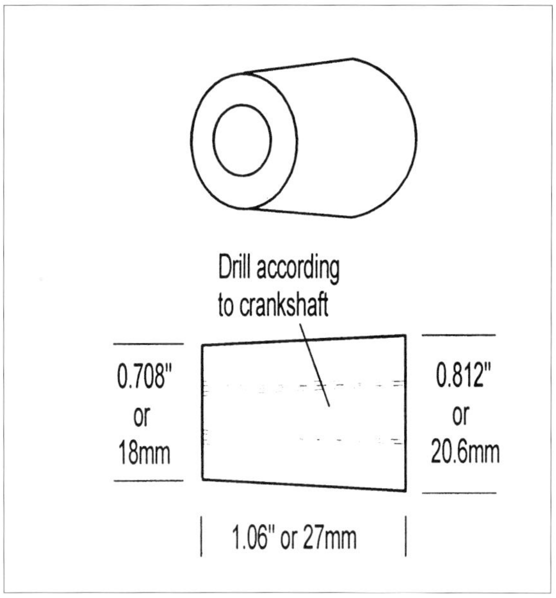

Fig. 9.13: Fillet (x 2) (Actual size)

Photo 32: Complete wooden structure of the Glasgow model replica

Chapter 10
Metal Components - Glasgow Model

Construction Sequence 11 – 15

11 Machining of flywheel,
12 Machining of crankshaft and bearing shells,
13 SUB-ASSEMBLY – stage 3,
14 Machining of power cylinder, pressure transfer pipe, power piston, clevis, con-rod and fittings, power crank and pin.
15 SUB-ASSEMBLY – stage 4.

Sequence 16 - 19

16 Water cooling jacket, displacer cylinder, sealing flange, displacer rod bush/gland, pressure pipe outlet,
17 Displacer, displacer rod and clevis,
18 Displacer crank, crank pin, con-rod big-end and con-rod,
19 FULL SCALE ASSEMBLY and TESTING – stage 5

11 FLYWHEEL

NOTE: The original flywheel was made of cast iron to a high degree of finish most probably by Robert's brother James, a master craftsman in foundry-work in his early engineering years and it is quite possible, even probable, that he had a hand in the casting of the flywheel/crankshaft as well as other components used in this model.

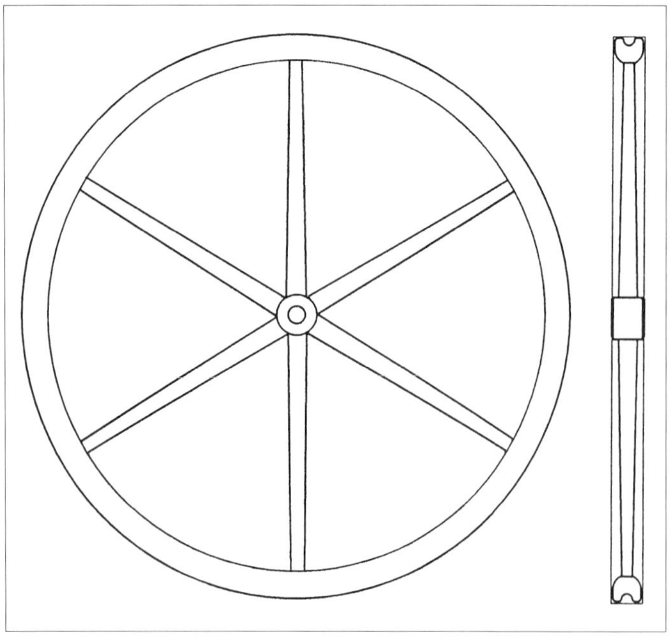

Fig. 10.1: Flywheel elevation and side view

The flywheel and crankshaft of the Glasgow model are cast in one piece, and unless one has access to a good pattern maker and foundry, it is difficult for the home machinist to reproduce this type of composite component. The author has chosen to make the two parts, flywheel and crankshaft, as separate items.

(Construction Note: Suggestions as how to make a similar flywheel: either make a pattern from ¾" plywood and have a casting made (aluminium is a good alternative to cast iron), or take an aluminium plate of suitable thickness, mark out the diameters and spokes, chain drill at the marked places, cut through, and file.)

The flywheel diameter is 12⁹⁄₃₂" = 12.28" or 312mm, rim thickness is ⅝+" = 0.63" (16mm) with a flat top and rounded underneath to a depth of ³³⁄₆₄–" = 0.511" (13mm), with a hub diameter of ⁶³⁄₆₄" = 0.984" (25mm) and six spokes approximately 5" (127mm) long, ¹⁵⁄₃₂+" = 0.472" (12mm) thick at the hub and ⁵⁄₁₆" = 0.311" (7.9mm) at the rim. The rim is grooved with a half-round depression of ³⁄₁₆+" = 0.196" (5mm) diameter which was meant to serve as a pulley, although there is no record of such use, (Fig. 10.1).

Since it is difficult to cast this flywheel and the crankshaft as one piece, the flywheel is bored and reamed ⅜" = 0.375" (9.5 or 10mm) to take a machined crankshaft.

12 CRANKSHAFT AND BEARING SHELLS

CRANKSHAFT

(NOTE: It is highly probable that the (cast) crankshaft tapers from where the fillets end towards the cranks; on the other hand the casting could have been made in such a way that the tapering starts from the flywheel hub. For some reason the position of the flywheel appears to be closer to the power side than to the displacer side, but this could also be due to various attempts to rehabilitate the engine. Since this is one grey area, unlikely to be resolved just as it unlikely that the model will be dismantled again, the author has assumed that the crankshaft shank is straight in the section of the flywheel and the fillets (total length given as 70mm or 2¾") and tapers from the fillets towards the ends.

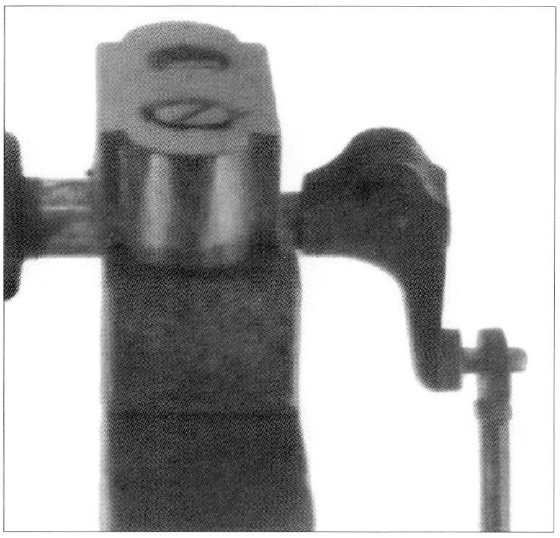

Photo 33: Close-up of the replacement cast crank of the Glasgow model

The crankshaft total length is $7\frac{51}{64}$" = 7.795" (198mm) with a straight shank for the length of the flywheel hub and the two fillets (given as 70mm or $2\frac{3}{4}$") and then tapering to the two ends as in Figure 10.2. Taking the centre of the flywheel thickness as the measuring point, the part of the crankshaft on the power cylinder side is calculated as $3\frac{21}{32}$+" = 3.66" (93mm) long, while the part from the mid-centre of the flywheel to the displacer cylinder end is $4\frac{1}{8}$+" = 4.133" (105mm).

From a close-up view of the crankshaft it appears that the displacer crank has been altered to the extent that the tapering crankshaft disappears inside the crank almost completely. (Photo 33).

BEARING SHELLS

(NOTE: Robert Stirling elected to insert bearing shells between the bearing brackets for the crankshaft to rotate with minimum friction. These shells were made from thin lead or copper sheets, pressed around the tapering ends of the crankshaft. Both the top and bottom shells had a wing-like extension on both sides of the crankshaft end so that the shells were retained in place by the pressure of the bearing brackets when bolted together.)

The bearing shells are cut from thin copper (or lead) sheet in two parts, pressed in a vice around the ends of the crankshaft and inserted between the bearing brackets which are marked and filed down to take the bearing shells flush to the surface of the brackets. (Figure 10.3).

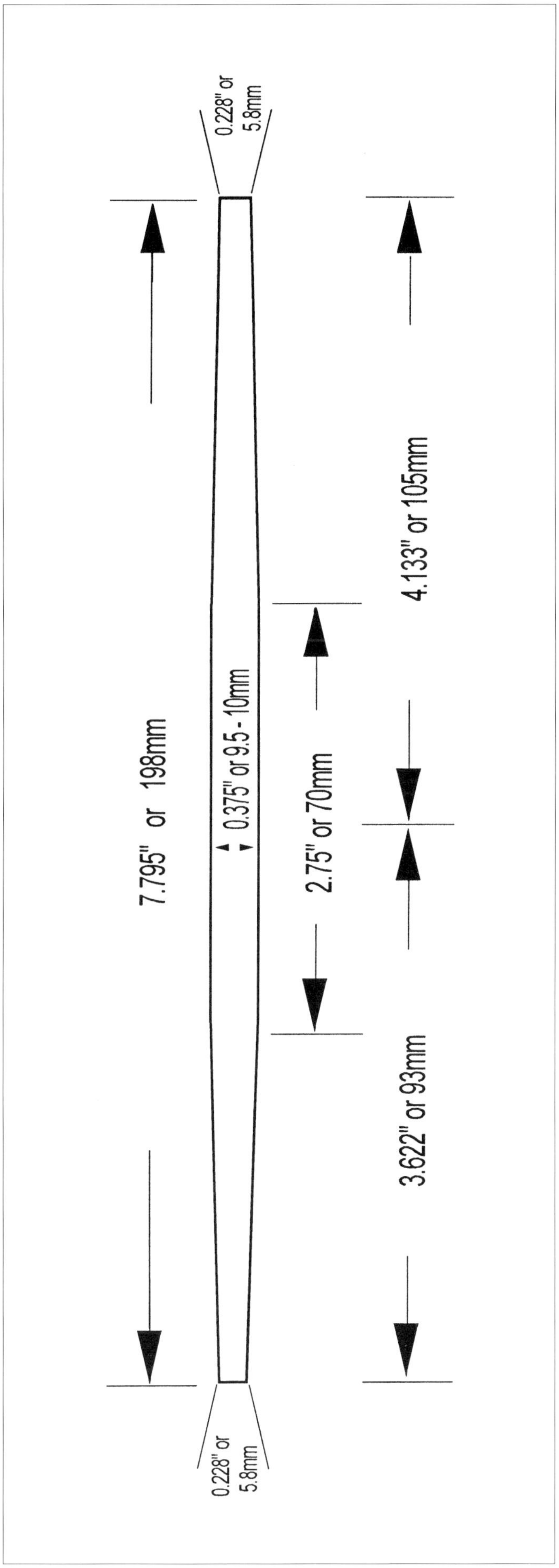

Fig. 10.2: Crankshaft with straight shank for the flywheel and fillets on either side, and tapering towards the ends. (Full size)

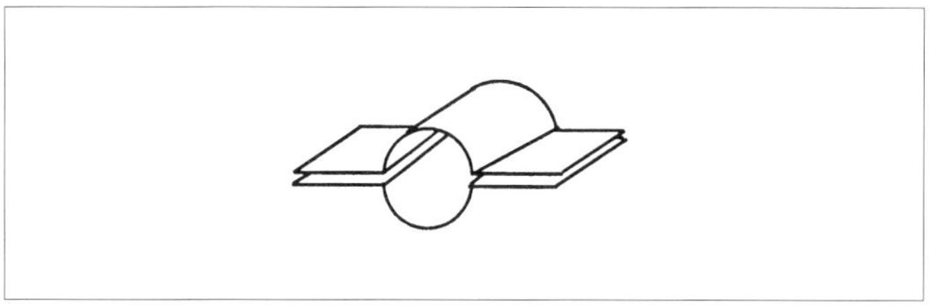

Fig. 10.3: Bearing shells

13 SUB-ASSEMBLY – Stage 3

The crankshaft is inserted into the flywheel hub and fillets initially with the centre of the flywheel hub in the centre of the crankshaft. Any adjustment to the position of the flywheel can be made at a later stage when the power and displacer cranks are in place. The displacer side of the crankshaft is slightly longer than the power side.

The crankshaft assembly is placed between the bearing shells on the bearing base and held down by the bearing caps, then placed on the lintels and screwed down with two countersunk 4mm x 50mm bolts. Any adjustments, filing or lubricating is done at this stage. (Figure 10.4 – Photo 34)

Photo 34: Sub-assembly of the flywheel on the top structure

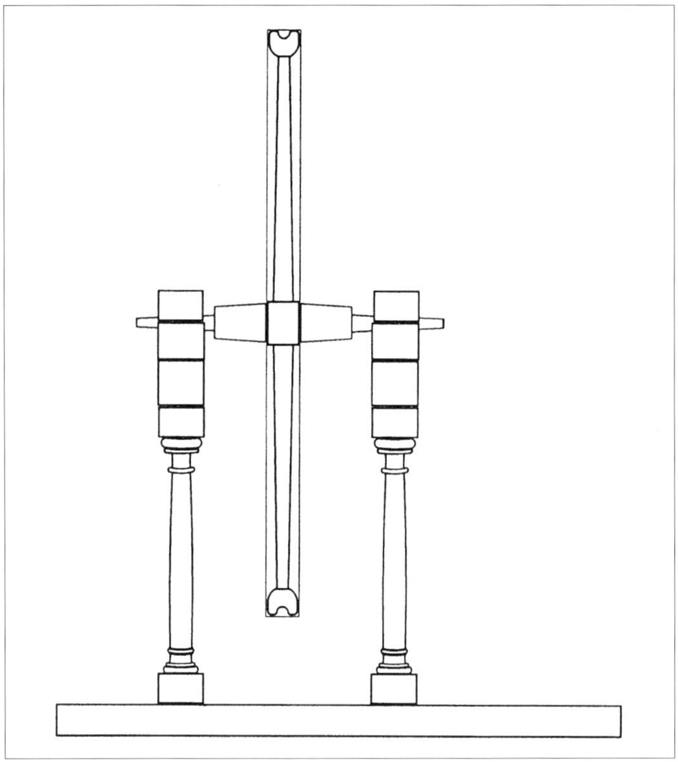

Fig. 10.4: Sub-assembly - Stage 3
Flywheel assembled on supporting brackets

14 POWER CYLINDER, PRESSURE TRANSFER PIPE, PISTON, CLEVIS, CON-ROD AND FITTINGS, POWER CRANK AND PIN

POWER CYLINDER

The power cylinder base has a pipe with a bend that forms part of the pressure transfer system. This could have been cast as one item originally. Since in this replica it is difficult to repeat the process of casting, the power cylinder and the pipe were machined separately and then soldered.

Readers are reminded about the sliding arrangement referred to in Section 9.

The specifications for the full size cylinder are as follows:

External Length:	3⁹⁄₆₄+"	= 3.15"	80mm
Internal Length:	3¹⁄₃₂"	= 3.03"	77mm
O.D.*	2⁵⁄₃₂+"	= 2.165"	55mm
I.D.	2³⁄₆₄"	= 2.047"	52mm

** The cylinder has a flange-type projection at the base with an O.D. of 2⁹⁄₃₂" = 2.283" (58mm), and ⅛" = 0.125" or 3.2mm thick.*

The cylinder is machined from a piece of cast iron bar 2²³⁄₆₄+" = 2.362" (60mm) faced and reduced externally and then bored, honed and lapped to give an internal diameter of 2³⁄₆₄" = 2.047" (52mm). (Figure 10.5) (Photo 35)

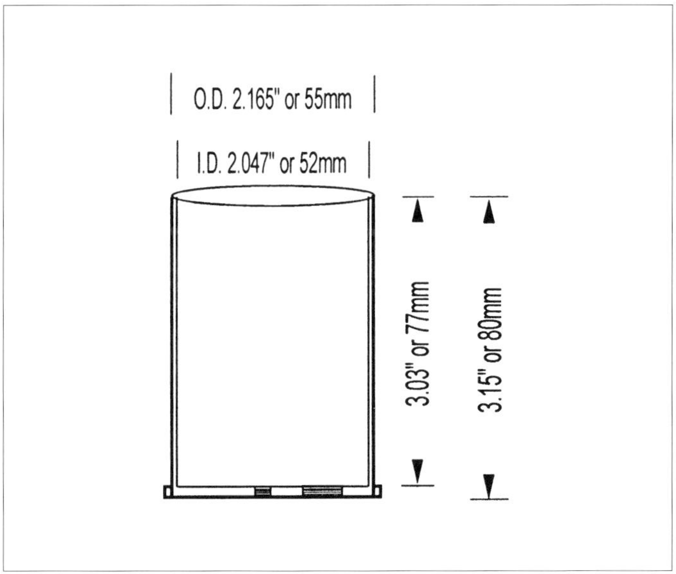

Fig. 10.5: Power cylinder

Photo 35: Power cylinder on its base (or 'in trial position on top structure')

While still in the chuck the base of the cylinder is centre-drilled, drilled and countersunk to take a 4mm bolt, and drilled 27/64" = 0.42" or 10.7mm between the centre hole and the inner wall to take a bend from the pressure (or communicating) pipe. (Figures 10.6a and 10.6b).

PRESSURE PIPE FITTING

NOTE: A common name for the pipe connection between the displacer cylinder and the power cylinder nowadays is 'pressure transfer pipe' or 'pressure pipe' or even simply 'connecting pipe'. Apparently documents held by the Hunterian Museum refer to this component as 'communicating' pipe which now relates more to inter-personal action than a mechanical one.

PRESSURE PIPE CONNECTING SYSTEM

It is difficult to discover the original layout of the pipe connection between the two cylinders of the Glasgow model and whether at one time this was similar to the Edinburgh one (which is the original). It could be that when the engine was rehabilitated a connector/socket and sleeve were added on at the power cylinder end or that these were shifted from the displacer cylinder end, or even that this was the original system.

We know for certain that when the engine was dismantled the complete power unit was made to slide in a slot on the engine platform, and therefore it was easier to connect the pressure pipe system at the power end. Figure 10.6a shows the author's interpretation of the connecting system and the various components. On the other hand while the Edinburgh model's connecting system was actually physically checked by the author the Glasgow model is in a narrow sealed showcase which makes it difficult to see those parts behind the columns or beneath the engine platform.

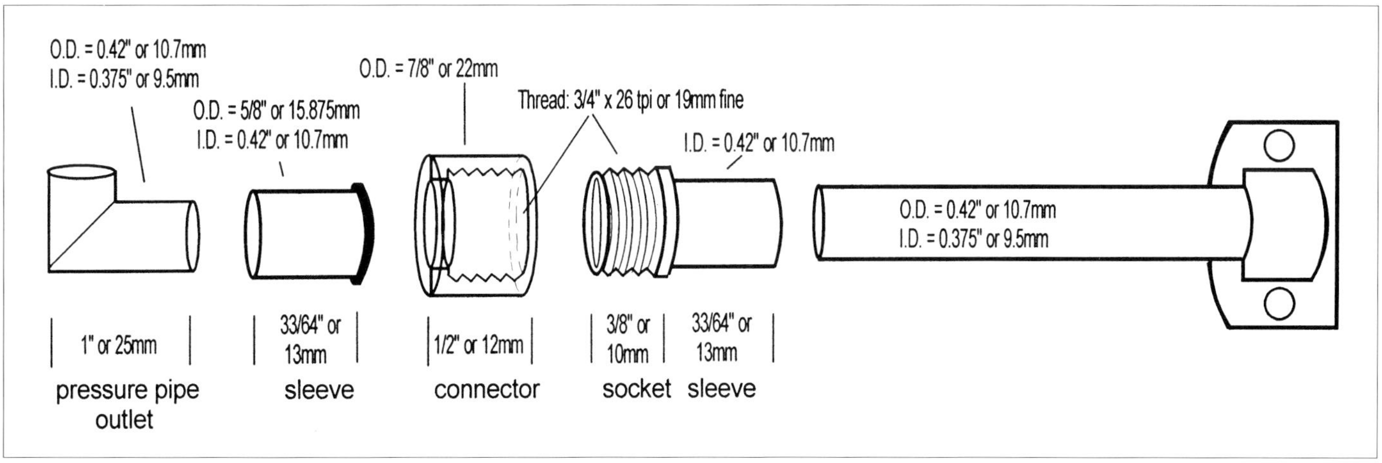

Fig. 10.6a: Layout of the pressure pipe connecting system (Not to scale)

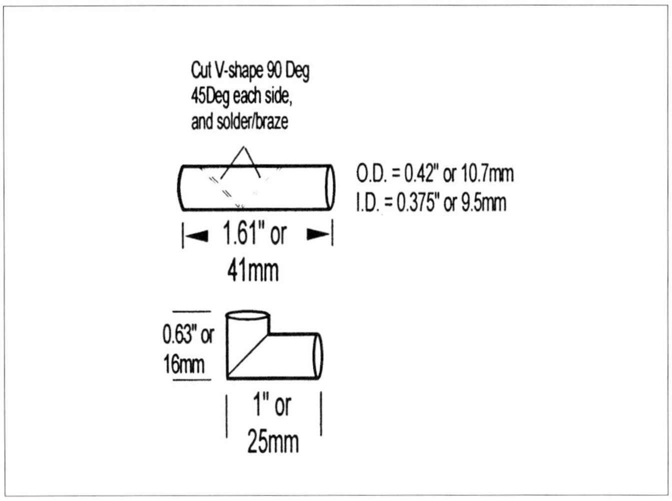

Fig. 10.6b: Pressure pipe inlet

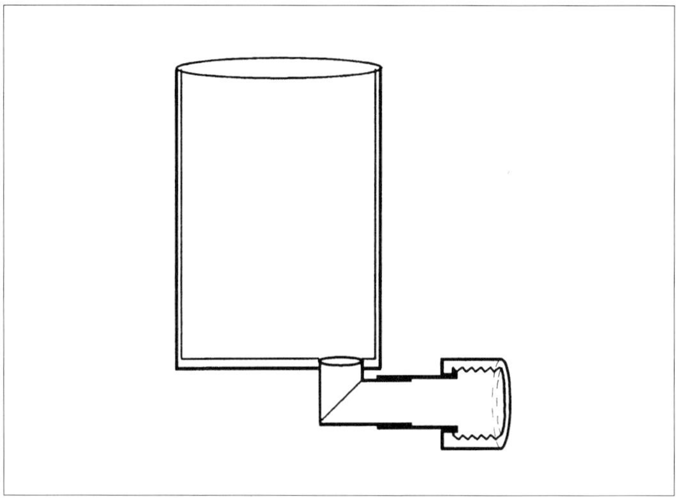

Fig. 10.6c: Pressure pipe sleeve/connector fitting

The connections can be divided into three sections, the pipe connected to the power cylinder, the pipe connected to the cylinder head of the displacer cylinder and the sleeve/socket/sleeve arrangement that brings the two pipes together.

The pressure pipe from the power cylinder has an O.D. of 27⁄64" = 0.42" (10.7mm), with an I.D. of ⅜" = 0.375" (9.5 or 10mm). A length of 1 39⁄64" = 1.61" (41mm) has a 90° wedge cut and soldered to leave a length of ⅝+" = 0.63" (16mm) for insertion into the power cylinder base as in Figure 10.6c.

A sleeve extension is machined from brass, O.D. ¾" (19.05mm), 19⁄32" = 0.59" (15mm) long, drilled to take the pressure pipe in a tight fit. The sleeve is then faced and externally reduced to ½" or 12mm for a length of 33⁄64" = 0.515" (13mm), leaving a 2mm ridge or lip (18mm) machined square. (Figure 10.6a).

The connector is a small fitting machined from solid brass rod O.D. 1" or 25mm and ½" or 12mm long, drilled through 15⁄32" = 0.47" or 12mm to take the sleeve body in a tight fit. The hole is then widened to I.D. 45⁄64" = 0.703" (18mm) for a length of 5⁄16" or 0.312" (8mm) and then threaded ¾" = 0.75" (19mm) × 26 t.p.i. or similar fine thread, with an undercut at the end. The end is finished square.

The connector and sleeve are fitted to the pipe from the power cylinder. (Figure 10.6c).

POWER PISTON

The Glasgow model has a working piston made up of 'two greased leather pads folded at the edges and sandwiched between two wooden discs – held on to the connecting rod by a brass

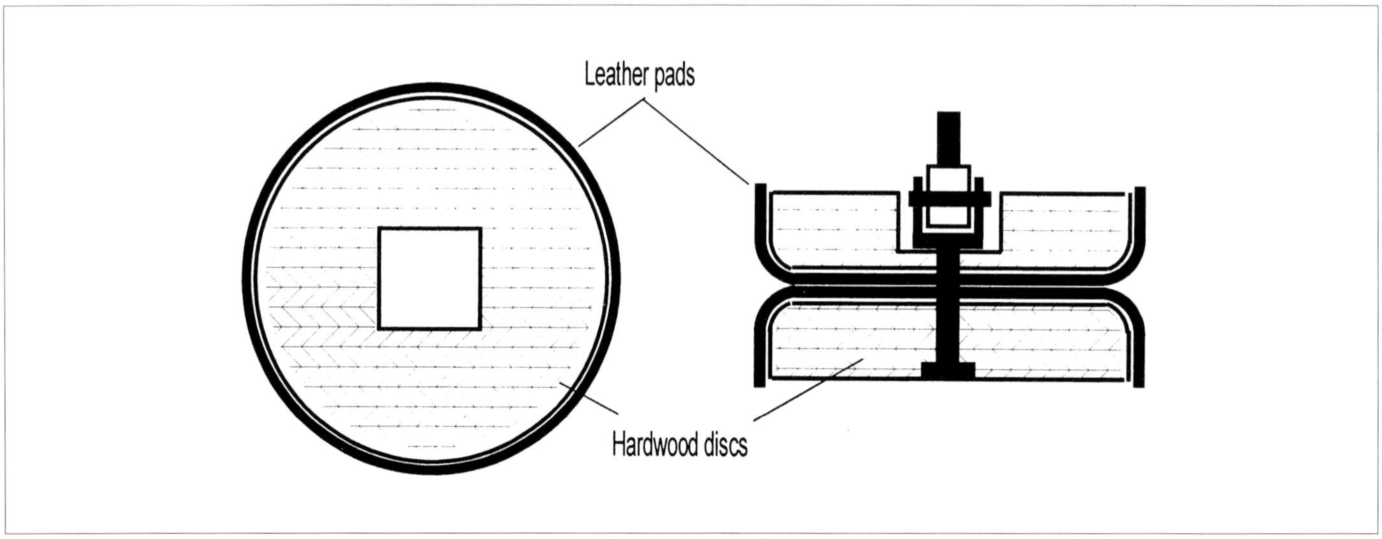

Fig. 10.7: Power (working) piston (Actual size)

nut' – description by Mike Jewkes of the Hunterian Museum. This tallies with the cut-away section of the replica held in the Science Museum, London. The detailed information about this component is very important as it throws light on how pistons were actually made in those days; also it gives the modeller a great opportunity to fabricate a piston to exact specifications. The method of making a double-sided leather piston has been described on Pages 37 and 38 (Chapter 7) and therefore there is no need to repeat the process.

The power piston (Figure 10.7) has a total diameter of $2\frac{3}{64}$" = 2.047" (52mm); assuming that the leather is $\frac{1}{16}$" or 1.5mm thick, therefore the hard wood discs have a diameter of $1\frac{59}{64}$+"= 1.93" (49mm) or slightly less and $\frac{3}{8}$" = 0.375" (9.525mm) thick. The discs are drilled and machined as for the Edinburgh model, with recess for the bolt head in the bottom disc and a larger recess in the top disc to take the clevis-type device.

As in the previous model the leather pads are dressed over the wooden discs, held by the clevis device in the form of a piston and are then inserted with a temporary con-rod into the working/ power cylinder with a few drops of light oil and worked up and down manually until the leather pads bed down.

CLEVIS

The same type of clevis-type device (Figure 10.8) that was used on the Edinburgh model is used to bolt the two hard wood discs of the leather piston together. This is made up of two parts, the bottom part is a bolt with a straight shank and threaded 8BA or 2mm for a length of $\frac{3}{32}$" = 0.093" or 2.5mm, while the top part is the actual clevis and is machined from $\frac{3}{8}$" = 0.375" or 10mm brass cube, milled $\frac{1}{4}$" or 6mm across and $\frac{19}{64}$" = 0.296" or 7.5mm deep, cross-drilled to take a 2mm pin. The centre of the hole is either 4.5mm from the bottom or 3mm from the top.

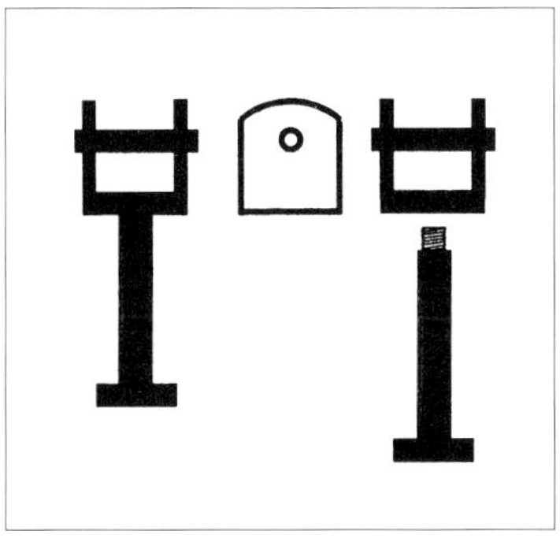

Fig. 10.8: Clevis-type device (Actual size)

CON-ROD SMALL-END

The small-end that fits into the slotted end of the power piston top wooden disc clevis is machined from $\frac{5}{16}$" = 0.312" (or 8mm) brass rod, $\frac{1}{4}$" or 6mm thick, centre-drilled and drilled $\frac{5}{64}$" or 2mm, then cross-drilled and tapped 6BA or 2.5mm for a depth of $\frac{1}{8}$" = 0.125" or 3mm to take the con-rod end. The small-end is filed down slightly on each side to allow free movement in the clevis top (Figure 10.9).

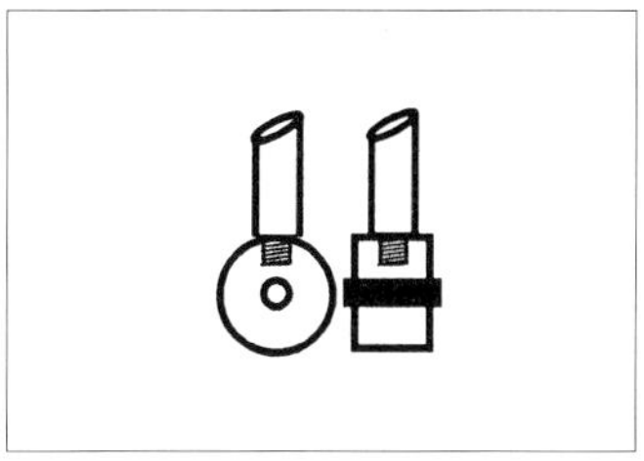

Fig. 10.9: Con-rod small-end (Actual size)

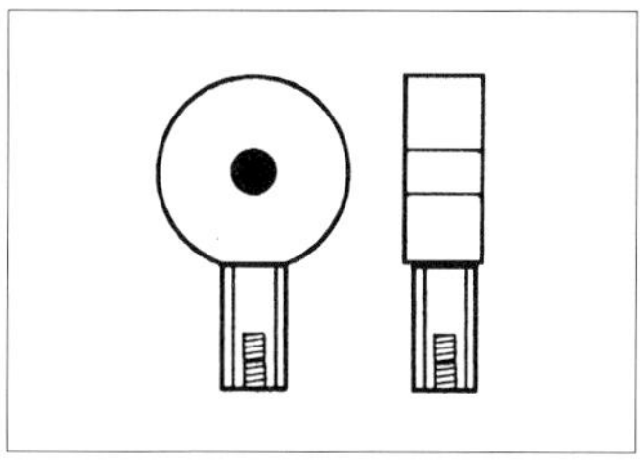

Fig. 10.10: Con-rod big-end (Actual size)

CON-ROD BIG-END

The big-end is machined from $\frac{3}{16}$–" = 0.180" (4.57mm) thick brass, total length (22mm) long, in two parts, a $\frac{17}{32}$" = 0.53" (13.5mm) round end, a narrow square neck $\frac{11}{32}$" = 0.343" (8.7mm) long. The top round end is centrally drilled to take a $\frac{1}{8}$" = 0.125" or 3mm crankpin, while the lower part is carefully drilled and tapped 6BA or 2.5mm to a depth of $\frac{5}{32}$" = 0.156" or 4mm to take the power con-rod. (Figure 10.10).

POWER CON-ROD

The power con-rod is cut from $\frac{9}{64}$" = 0.140" or 3.5mm (or $\frac{1}{8}$" = 3.2mm) mild steel rod approximately $4\frac{1}{64}$" = 4.015" (102mm) long, threaded both ends 6BA or 2.5mm. When assembled the distance from mid centre of the big-end to mid-centre of the small-end is $4\frac{31}{64}$" = 4.488" (114mm), (Figure 10.11).

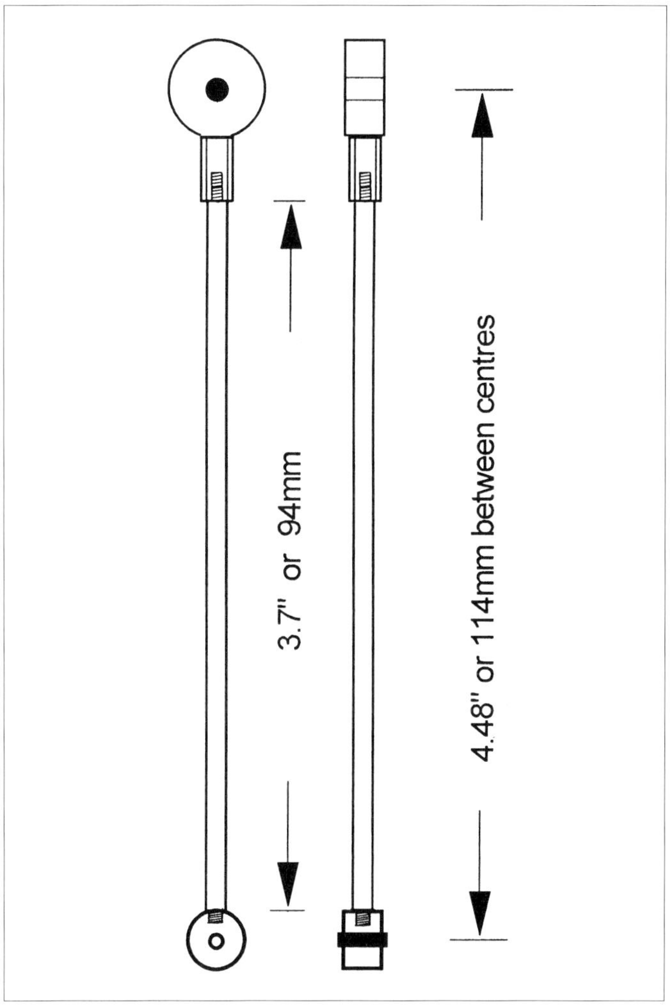

Fig. 10.11: Power con-rod (Actual size)

Readers should however check the actual length of the rod when the crankpin is at its lowest point and the power piston just above the base of the power cylinder. Any fine adjustment in the length of the con-rod can be done with screwing in or out the ends.

POWER CRANK

NOTE: There is a divergence of opinion vs. given data about the length of stroke of the power piston. The author was given 24mm (15/16+" = 0.945mm) as the full stroke, whereas by measurement and extrapolation the stroke was found to be closer to 40mm (1 37/64-" = 1.574"). A close-up image of this component with the flywheel brackets in the background backs this assumption – a line drawing depicting the exact position of the power crank in the foreground of the brackets shows the reasoning behind the 40mm stroke – Figure 10.12). Later when examining some papers in the archives of the Museum a roughly drawn drawing with notations made when the engine was dismantled for repairs describes the stroke as 1 21/32" = 1.65" (42mm).

For those machinists who prefer to keep to the original version as machined by Robert Stirling, the cast crank can be replaced by a machined crank similar to that of the Edinburgh model which is definitely the original crank. This is cut and filed as in Figure 12.13b with a 1/8" mild steel pin inserted at a distance of 53/64" = 0.826" or 21mm.

The power crank appears to be a cast item. The actual profile of this replacement, and extension to the crankshaft, is shown in Figure 10.13. The total length of the crank is 1 11/64+" = 1.18" (30mm) long, 13/64-" = 0.200" (5.08mm) thick at the crankshaft end, tapering to 5/32+" = 0.163" (4.15mm) at the other end, 7/16" = 0.437" (11.11mm) wide at the top tapering to 21/64" = 0.326" (8.3mm) before rounded off.

The crankpin-retaining projection which is part of the casting has a diameter of 13/64" = 0.203" (5.16mm) and is 9/32" = 0.281" (7.14mm) long, drilled to take a pin (the actual crankpin) O.D. 1/8" = 0.125" or 3.2mm and 3/8" = 0.375" or 9.5mm long. The distance between the centres of the crankshaft end and of the crankpin is 53/64" = 0.826" or 21mm.

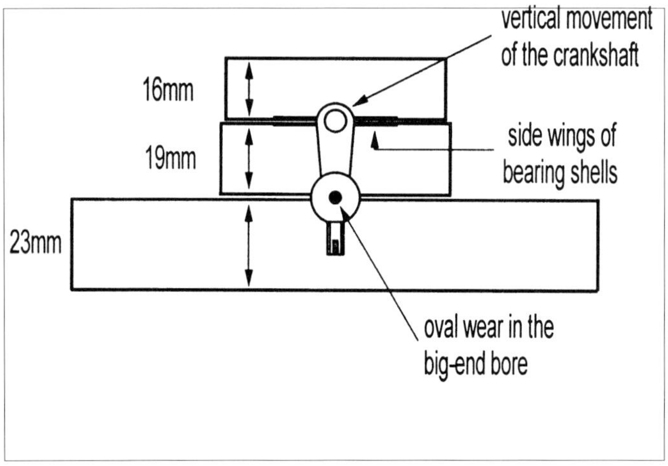

Fig. 10.12: Diagram representing the position and size of the power crank in relation to the bearing brackets/lintel

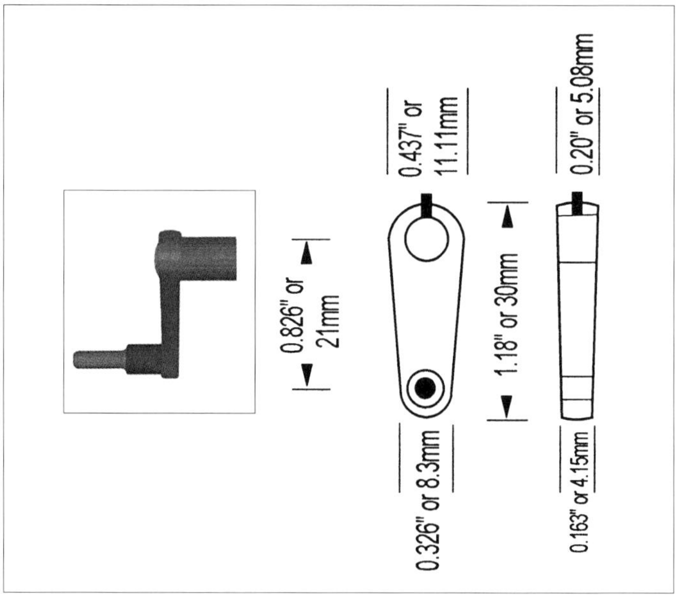

Fig. 10.13: Power crank specifications (Inset - a cast crank)

At the top end the drilled hole for the crankshaft end tapers slightly from 7/32+" = 0.228" (5.8mm) to 15/64" = 0.234" (5.95mm) as in Figure 10.12. The crank is then cross-drilled with a 2mm drill at the point where it is to be mounted on the crankshaft. *(Later on sub-assembly the crank is pinned to the crankshaft end.)*

15 SUB-ASSEMBLY – STAGE 4

(Temporary assembly) The power cylinder and wooden base are temporarily screwed to the engine platform**, and the power piston and con-rod inserted into the cylinder. The crank is mounted on the crankshaft, the con-rod big-end is inserted on the crankpin and secured lightly with the retaining pin. A few turns of the flywheel will show whether both the stroke and the piston fitting work well. In fact if the pressure transfer hole is stopped by a finger, the piston should not budge. (Figure 10.14) (Photo 36).

The fitting of the power cylinder and its base to the original Glasgow model was made through a slot in the base which allowed the power cylinder/base to be moved to the left (backwards –* ***see note on 'Pressure Pipe Fitting' above).

It has been suggested, (and almost definitely correct), that this was done during the rehabilitation period as it is probable that various components were taken out, any soldering removed and parts such as the socket/sleeve and the screw/washer fitting added on, while the slot in the engine platform must have been made at the same time.

The author suggests that there is no need for the slot/sliding arrangement; some machinists may for historical purposes include this arrangement although it is almost definitely an add-on to the original model. Others may wish to keep what may have been Robert Stirling's own model before rehabilitation.

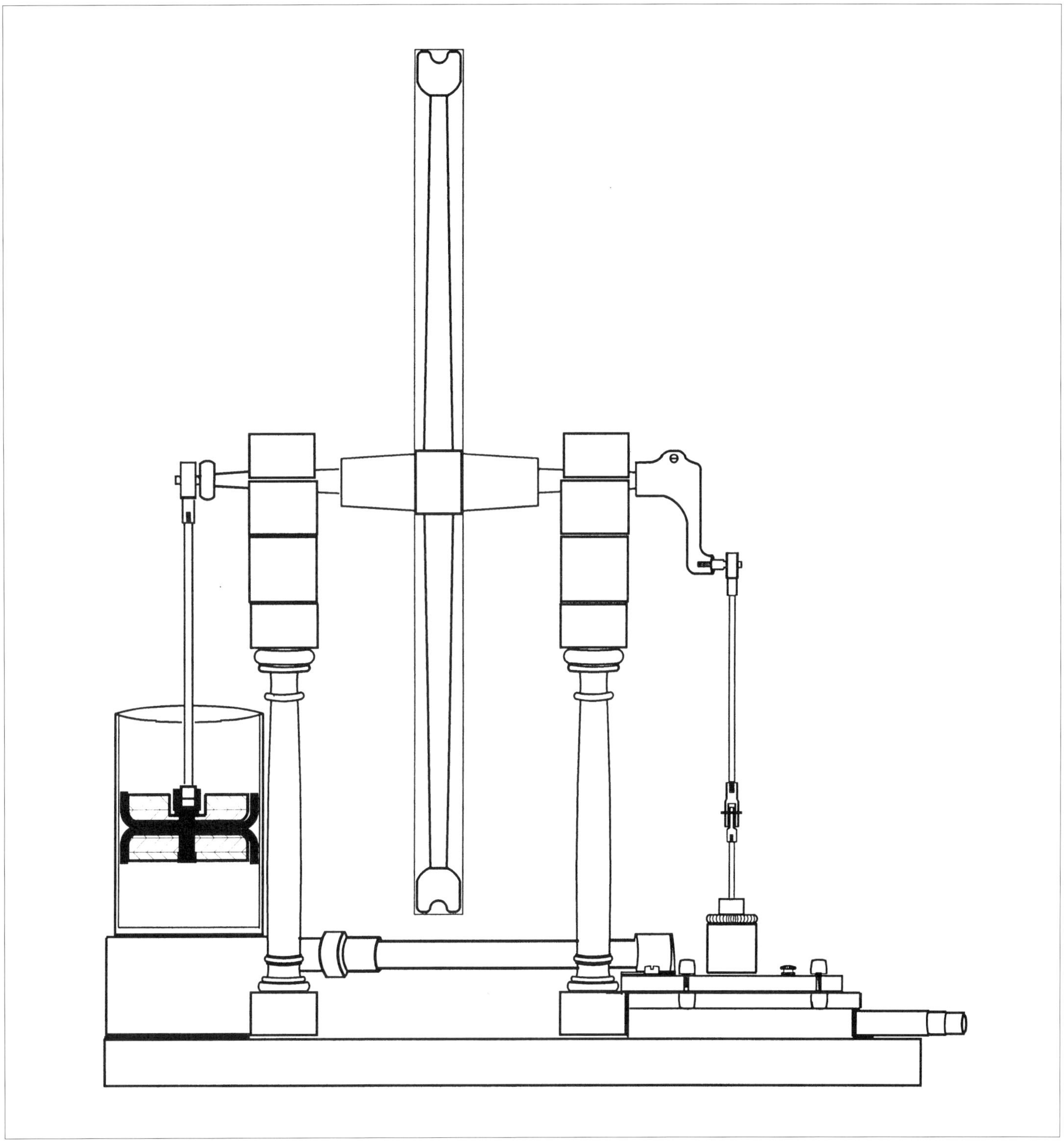

Fig. 10.14: Layout of the drive mechanism and the connecting pipe between the two cylinders

DISPLACER UNIT

Sequence nos. 16 - 19

16 Water cooling jacket, displacer cylinder, sealing flange, displacer rod bush/gland, pressure pipe outlet,

17 Displacer, displacer rod and clevis,

18 Displacer crank, crank pin, con-rod big-end, con-rod small-end and displacer-con-rod,

19 FULL SCALE ASSEMBLY and TESTING – stage 5.

16 WATER COOLING JACKET

(NOTE: The addition of a water cooling jacket on the displacer cylinder has given rise to speculation as to whether this was on the original model, or an add-on when the engine was rehabilitated some 20 years after Robert Stirling donated it to the University of Glasgow in 1826.

As explained earlier, Robert Stirling could have realized, after testing the first model, particularly as the displacer cylinder was made of brass, that heat rises quickly along the length of the

Photo 36: Power unit coupled to the crankshaft

cylinder and that therefore some sort of buffer was necessary to control this vertical conduction of heat and to keep the top part as cool as possible, thus maintaining the temperature difference necessary to run an engine efficiently. The construction of the water jacket was such that the water inside was open to the atmosphere and could therefore also dissipate heat in this manner.

The other school of thought maintains that the water jacket was an add-on when the engine was being rehabilitated as the problem surfaced with decreased efficiency after the engine had been running for some time. This is supported by a statement made by Prof. Thomson ".... It is expedient to have a stream of water rushing over the upper part, thus carrying away the heat".)

The cooling jacket or cylinder is machined from rolled thin mild steel sheet with a rolled lip and base, (Figure 10.15). The total length of the cylinder is 5⁵⁄₃₂" = 5.157" (131mm), O.D. 4²¹⁄₆₄" = 4.33" (110mm), I.D. 4¼" = 4.25" (108mm).

The jacket also has two short pipes, at the top (water outlet) and bottom (water inlet) of the cylinder, obviously meant to be connected to filling/draining system. The original pipes are stepped and the end has an O.D. of ½" (12.7mm). Two brackets soldered to the jacket to secure the displacer unit to the engine platform. The brackets are cut from 5⁄64" = 0.078" (2mm) flat bar, mild steel or brass, each 1²⁹⁄₆₄+" = 1.456" (37mm) long and ½" (12 – 12.5mm) wide, bent at right angle, with the vertical part being ⅝+" = 0.63" (16mm) long, and the horizontal part being 53⁄64" = 0.826" (21mm) long, drilled to take two 4mm screws.

The brackets are placed at approximately 120° facing outwards (i.e. to the right).

DISPLACER CYLINDER (CAD Drawing 3)

NOTE: There is some difficulty in correctly identifying the metal of the displacer cylinder. Mike Jewkes of the Hunterian Musem, in his description to the author, stated that the displacer cylinder was made of cast iron. A paper with some roughly drawn drawings found among the documents in the archives of the Museum indicate that it is a brass cylinder – all other details of length and outside/ inside diameter are correct and tally with the Curator's details.

From close viewing of the model, encased as it was, it appeared that the displacer cylinder was made of rolled sheet iron as can be seen in Photo 37. At most it could be a rolled brass sheet cylinder. The bottom is definitely soldered and there appears to be a hint of rust. The cylinder has a wall and base thickness of ⅛-" – 0.118" (3mm), and 4 lugs protruding around the top circumference to take fasteners from the cylinder head.

The displacer cylinder is best machined from mild steel pipe with a wall thickness of ⅛" (3.2mm) or metric 3mm thickness. The closest size of ready-made pipes are those used in the

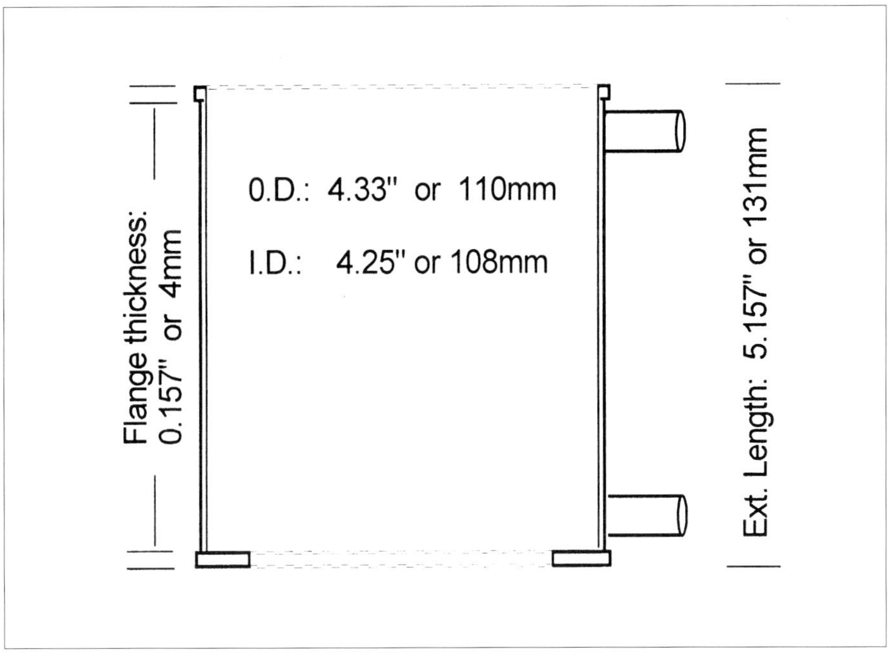

Fig. 10.15:
Water cooling jacket

manufacture or repair of car silencers (mufflers) which in my country vary from 2mm, 2.5mm and 3mm in wall thickness, depending on country of origin. $3\frac{1}{4}$" (82.55mm) pipes are a standard size with a wall thickness of $\frac{3}{32}$+" (actual thickness = 0.0984" or 2.5mm). *(From experience gained in obtaining off-cuts to make my own cylinders, these types of pipes are usually made in sizes that fit into each other for welding, e.g. a 3" pipe will fit into a $3\frac{1}{4}$", which in turn fits into a $3\frac{1}{2}$"- in these sizes the wall thickness is almost uniform to make it easier for the welding process. Other pipes received recently by the operators of these workshops were in metric sizes such as 30mm, 35mm, 40mm etc.) (Photos 38 and 39).*

The displacer cylinder has an external length of $10\frac{1}{2}$" = 10.51" (267mm) and an internal length of $10\frac{25}{64}$" = 10.39" (264mm). The outside diameter is $3\frac{7}{32}$+" = 3.228" (82mm) while the internal diameter is $2\frac{63}{64}$+" = 2.992" (76mm), (Figure 10.16). The critical measurements in order to obtain the same volume displacement are the internal diameter and the internal length.

The lugs mentioned above can either be machined from brass and brazed on to the top edge of the cylinder or machined/ fabricated from a ring-type flange which is then brazed to the cylinder. This latter method was used in the half-size replica.

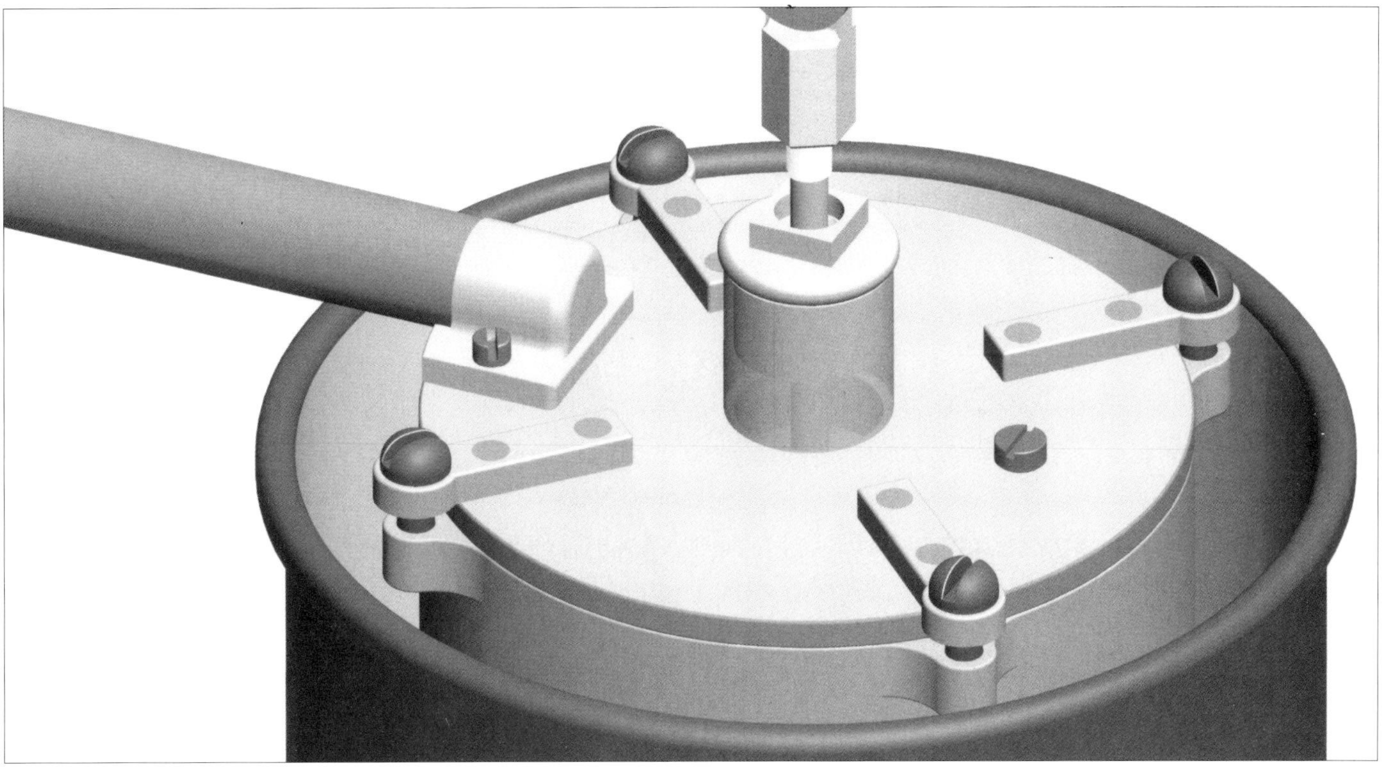

Illustration 3: CAD drawing showing the displacer cylinder in the water cooling jacket. (Courtesy - Keith Strong)

Photo 37: Close-up view showing the seams of the displacer cylinder and the water-cooling jacket

Photo 38: Bright mild steel pipe used to make a similar displacer cylinder

The lugs project by about $\frac{17}{64}$+" = 0.275" or 7mm, are 6mm half-round and $\frac{3}{16}$" = 0.1875" (4.7mm) thick, drilled and tapped to take 8 BA or 2.5mm bolts. Although this is a demanding job, in actual fact this is the closest one can get to the original model.

Alternatively a ring type flange can be machined from a 4" or 100mm brass round bar, reduced in diameter to $3\frac{25}{32}$" = 3.78" (96mm), $\frac{1}{4}$" or 6mm thick and bored to give a precision fit to the cylinder (82mm), then marked as in Figure 10.17 where the lugs are placed, and with great precision either milled or filed down to leave a ring $\frac{1}{8}$" = 0.125" or 3mm in depth. This latter alternative does not look as good on the appearance of the cylinder as the original, but when the cylinder head is screwed on the displacer cylinder, the effect is quite unobtrusive.

Photo 39: Various sizes of bms pipes suitable for hot air engines

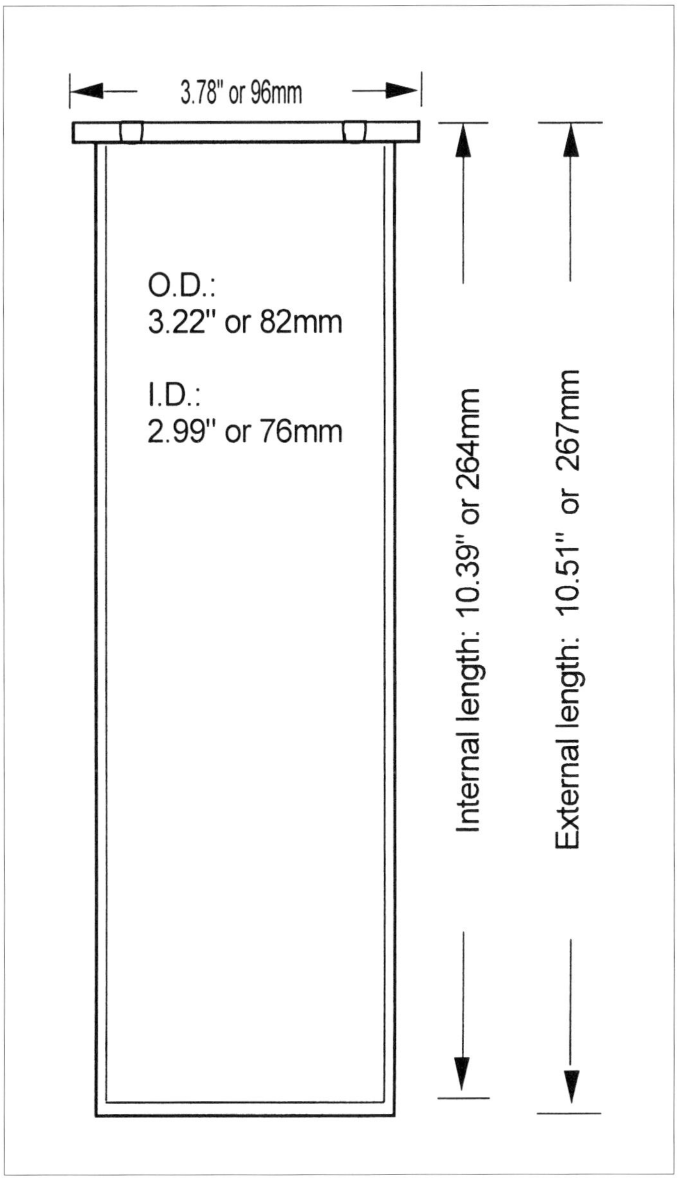

Fig. 10.16: Displacer cylinder

CYLINDER HEAD

The cylinder head of the original model was also cast, probably in bronze, and has projecting lugs that conform to the shape, size and to the distance between each projection on the displacer cylinder. A close-up image of the top also shows that these lugs were riveted to the surface.

The casting can be replaced by a machined component from brass round bar to the same diameter of either the displacer cylinder, or if a ring-type external flange is machined and fitted, to the diameter of the ring, excluding the projecting lugs. The finished casting has the following thickness: 15⁄64" = 0.234" (5.95m or 6mm), the part that goes into the cylinder reduced in thickness to 5⁄64" = 0.781" or 2mm to a diameter of 3" = 2.99" (96mm) to give a tight fit. The diameter of the cylinder head excluding the lugs is that of the displacer cylinder flange.

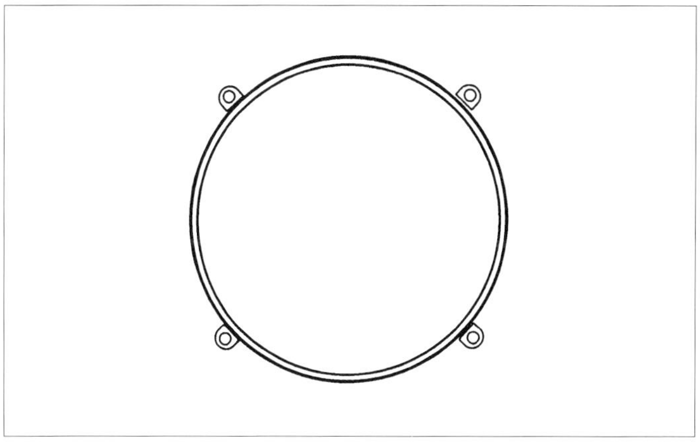

Fig. 10.17: Position of bottom lugs on displacer cylinder flange

The lugs are machined from brass flat, each 5⁄8" = 0.625" or 16mm long and 6mm wide, with a narrow tail and a semi-circular projection, (see profile inset in Fig.10.18), then drilled to take clear 8 BA or 2.5mm bolts that bolt the lugs to the displacer cylinder. The lugs are either screwed on or riveted to the cylinder head. (Photo 40).

The cylinder head is additionally centre-drilled, drilled and tapped or threaded BS Fine 9⁄16" or alternatively 15mm x 1 or 1.5mm – in any case an identical thread to the displacer rod bush or gland mentioned below. A side elevation of the sealing flange is seen in Figure 10.19.

There are two more operations to be done – drilling a hole for the pressure pipe outlet vent; this can only be done when the pipe outlet has been constructed and placed on the flange in the right direction in relation to the lugs, and drilling/tapping a pressure release vent 0 BA or 6mm on the opposite side to the pressure pipe outlet.

A decorative screw with a knurled top (inset in Figure 10.18) is machined and threaded 0 BA or 6mm to close the pressure release vent when the engine is prepared to run. A small 'O' ring ensures that no pressure escapes from this vent.

Displacer rod guide bush (or gland)

The displacer rod guide bush of the Glasgow model is machined in almost every way as the Edinburgh model with the exception of the top of the upper section which ends in a square shape nut instead of an octagon shape. The dimensions are slightly different as can be seen in Figure 10.20: while the diameter of the bush is the same, the length of the lower section is longer by about 0.10" or 2.5mm.

As in the previous model, the two sections have an identical thread both for insertion into each other and for mounting on the top flange.

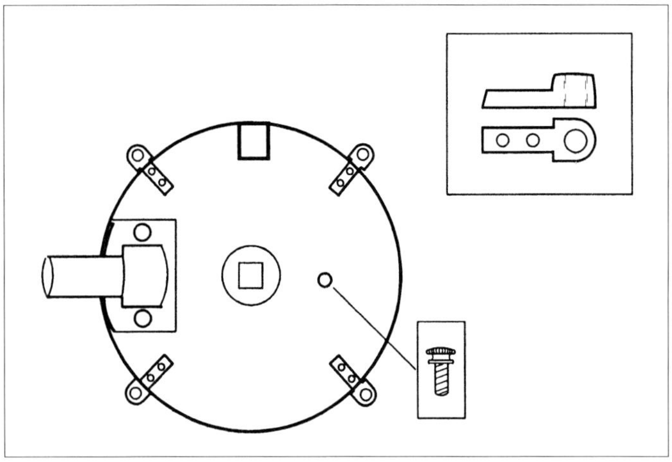

Fig. 10.18: Cylinder head/flange with position of lugs, pressure pipe outlet and displacer rod gland

(Insets - both Actual size:

Top right: profile of lugs, side elevation and top view

Bottom right: decorative screw for sealing pressure vent)

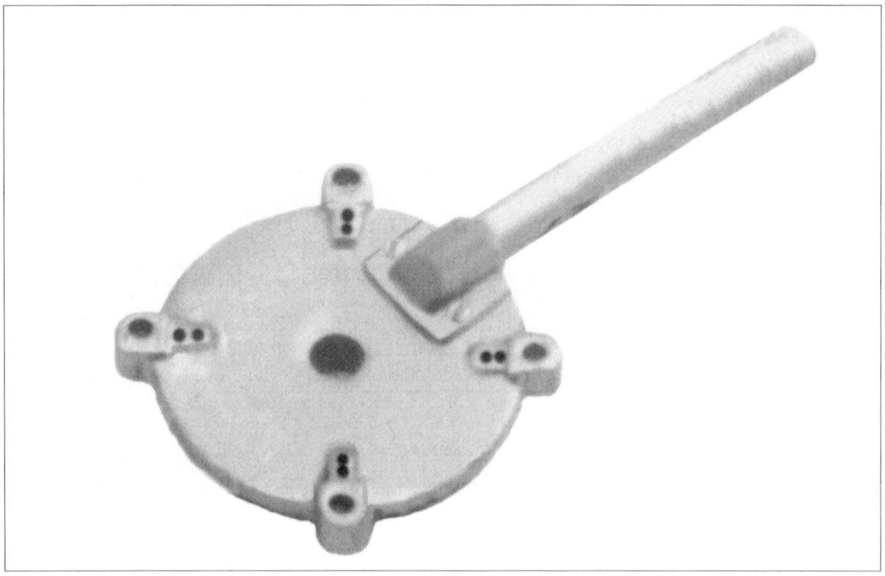

Photo 40:
Displacer cylinder head with pressure pipe

PRESSURE PIPE OUTLET

While there can be some slight doubt whether the pressure pipe outlet in the Edinburgh model was cast as part of the cylinder head – due to the way it was bolted to the displacer cylinder, in the case of the Glasgow model the pipe outlet was a separate cast item and bolted separately to the cylinder head, well inside the diameter of the remaining bolts around the circumference. Figure 10.18.

A casting for the pipe outlet is not absolutely essential as with some patience and precision bench-work it can be readily reproduced with two different diameters of brass pipe and a brass flat bar. The three parts are: a length of pipe O.D. $^{27}/_{64}$" = 0.42" (10.7mm), I.D. $^{3}/_{8}$" = 0.375" (9.5mm) and $4^{23}/_{32}$" = 4.72" (120mm) long, a short piece of $^{5}/_{8}$" = 0.625" (15.875mm) O.D. bar, $^{7}/_{16}$" = 0.437" (11.11mm) long and a flat bar $^{5}/_{64}$" = 0.078" (2mm) thick, $1^{11}/_{64}$" = 1.18" (30mm) wide and $^{25}/_{32}$+" = 0.787" (20mm) long cut to shape as in Figure 10.21.

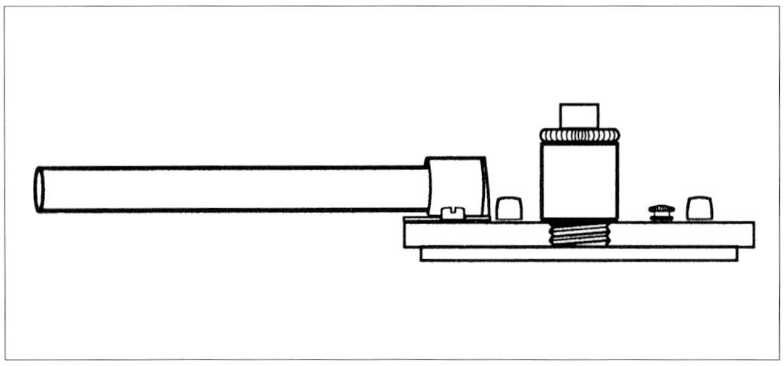

Fig. 10.19: Side elevation of sealing flange with lugs, pressure pipe outlet and displacer rod gland

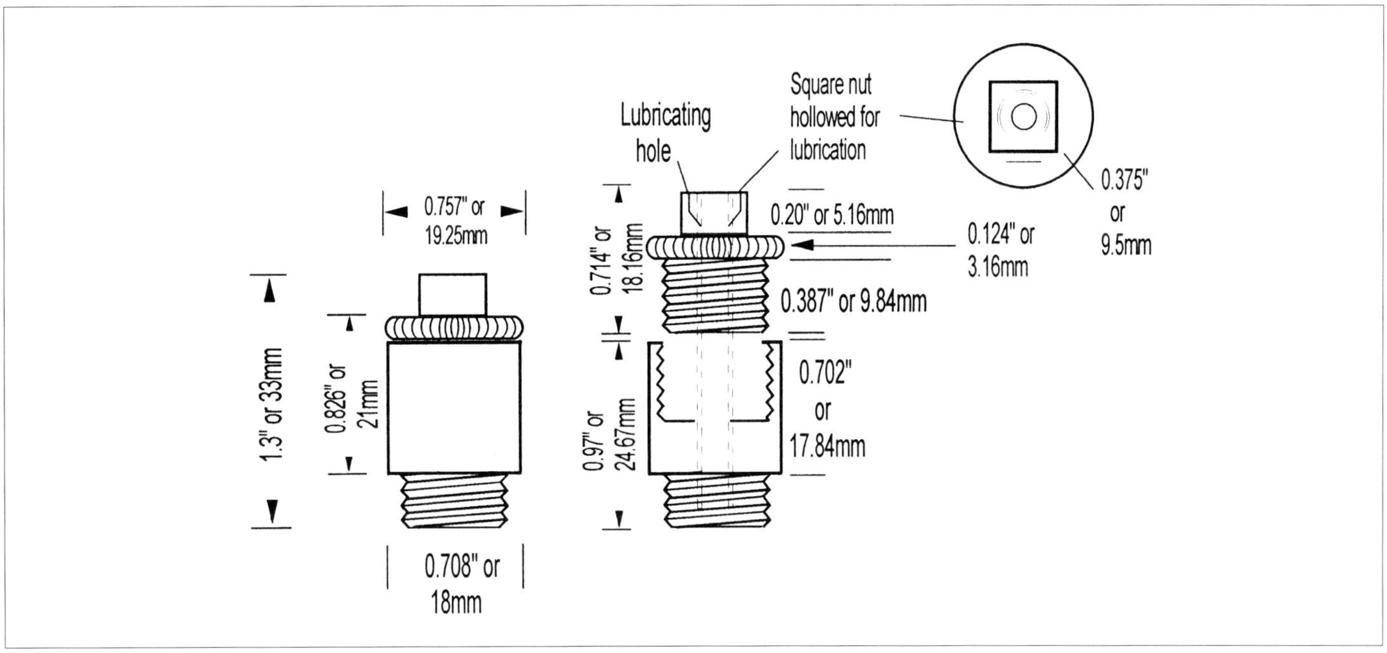

Fig. 10.20: Displacer rod gland/bush (Actual size)

The 5/8" round bar is first drilled to take the long pipe to a depth of 11/32" = 0.34" (8.7 or 9mm), then filed down on one side to give a flat base. The 12mm pipe is inserted and sweat-soldered, and the piece is then soldered to the flat bar drilled to take screws for bolting to the cylinder head. At a later stage this outlet is drilled from underneath through the sealing flange with a 11/32" or 8.5mm drill to provide an opening for pressure from the displacer cylinder to the power cylinder.

The pipe outlet can now be located on the cylinder head and the outline of the pipe outlet and the drilled holes marked on it. The pipe outlet is either screwed down to the cylinder head with a thin layer of epoxy adhesive or screwed down and soldered all round the base to ensure that there is no escape of air pressure.

17 DISPLACER, DISPLACER ROD AND CLEVIS

DISPLACER

(NOTE: From details given to the author, the displacer of the Glasgow model was made "…..from very thin copper sheet – thickness unknown". The machinist has four options for this component: 1. use thin copper sheet, rolled into a cylinder of appropriate measurement, sealed top and bottom; 2. find a suitable copper pipe and machine it to the specifications given below; 3. find a suitable brass pipe or brass sheet and finish it as above or 4. find a bright mild steel pipe, reduce wall thickness to 1/32" = 0.04mm or 0.8mm, even thinner if possible to produce a very light weight displacer.)

The displacer is 7 33/64+" = 7.52" (191mm) long, with an O.D. of 2 7/8" = 2.87" (73mm) – *given dimensions*, (Figure 10.22). If modeller is using any of the first three options for constructing this type of displacer, an alternative method to sealing the cylinder by soldering, is to use a hollowed aluminium sealing plug, with a narrow wall and a thin crown, and a button-type protrusion in the middle to take the displacer rod, suitably threaded 6 BA or 3mm, (Figure 10.23). An 8 BA or 2mm hole to one side allows the plug to be inserted tight fit with a smear of *Super Steel adhesive (by Plastic Padding).* The hole is sealed with a short 8 BA or 2mm bolt with adhesive. (Photo 41).

DISPLACER ROD

The displacer rod is cut from bright mild steel 1/8" = 0.175" (3.2mm) or 9/64" = 0.14" (3.5mm), 5 3/16+" = 5.196" (132mm) total length, threaded both ends 6 BA or 3mm for a length of 5/32" = 0.156" or 4mm, leaving a length of 4 7/8" = 4.88" (124mm) un-machined. One end goes into the clevis bottom fitting while the other end goes into the displacer top. (Figure 10.24).

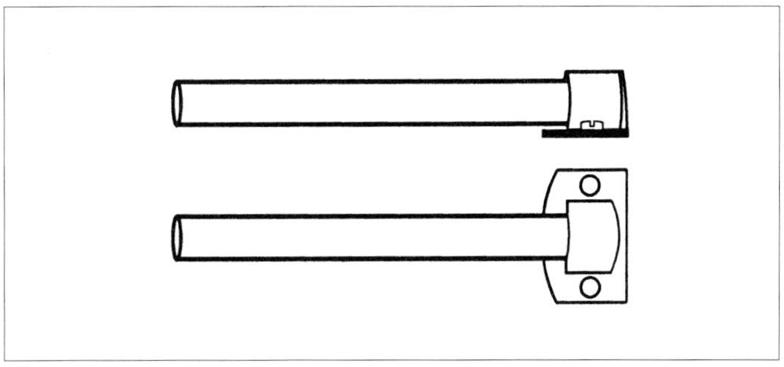

Fig. 10.21: Pressure pipe outlet

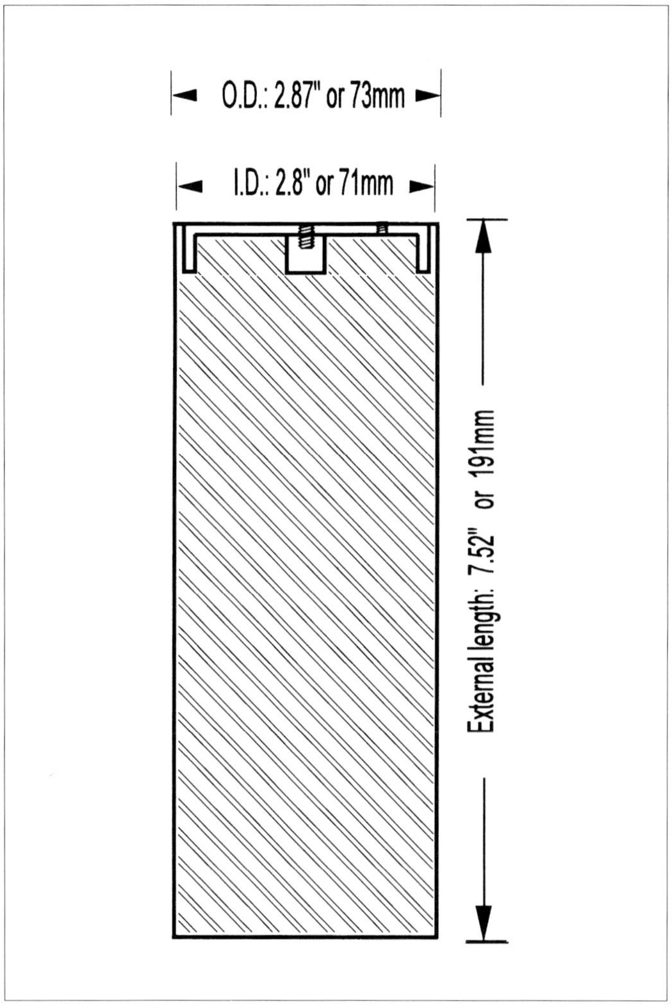

Fig. 10.22: Displacer

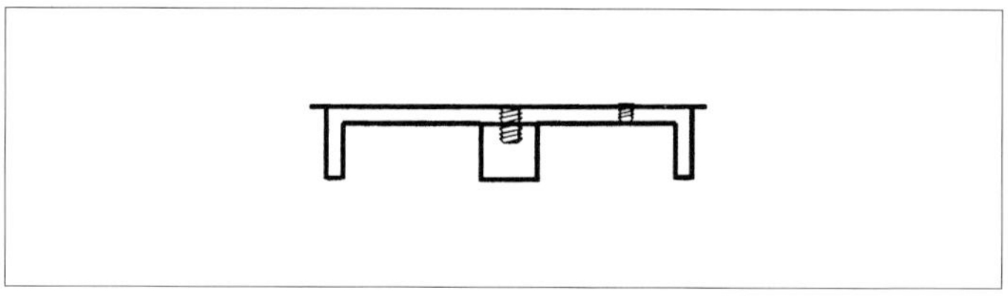

Fig. 10.23: Sealing plug

CLEVIS

The clevis is a two part component with a decorative top part, machined from brass, 5/64" = 0.20" (5.16mm) thick. Both parts are 5/8" = 0.625" (15.875mm) long, (Figure 10.25). The top end rounded part is 27/64" = 0.042" (10.7mm) wide, narrowing to 7/32" = 0.218" or 5.5mm neck, widening to 5/16" = 0.312" or 8mm before closing down to 11/64" = 0.17" (4.36 or 4.37mm). The thickness of the brass is reduced from the neck upwards to end up square at the top (11/64").

The clevis top end is cross-drilled to take a 1/16" = 0.0625" or 1.5mm pin in the centre of the round part, then drilled and threaded from the top 6 BA or 3mm to take the displacer con-rod, and finally slot-drilled or filed 5/64" = 0.078" or 2mm to take the bottom part of the clevis.

The clevis bottom end which is also machined from 13/64" = 0.20" (5.16mm) thick brass, has the top part reduced in thickness to 5/64–" = 0.078" (1.8mm actual) and filed round with a diameter of 3/8" = 0.375" or 9.5mm, formed like a neck but only 5/16" = 0.312" or 8mm long (slightly oval). This part rotates in the clevis and is cross-drilled just short of the centre to allow free movement in the clevis top. The bottom part is filed 1 1/64" = 0.17" (4.37mm) square and rounded near the end. The clevis bottom end in then drilled and tapped 6 BA or 3mm to take the displacer rod. (Figure 10.24).

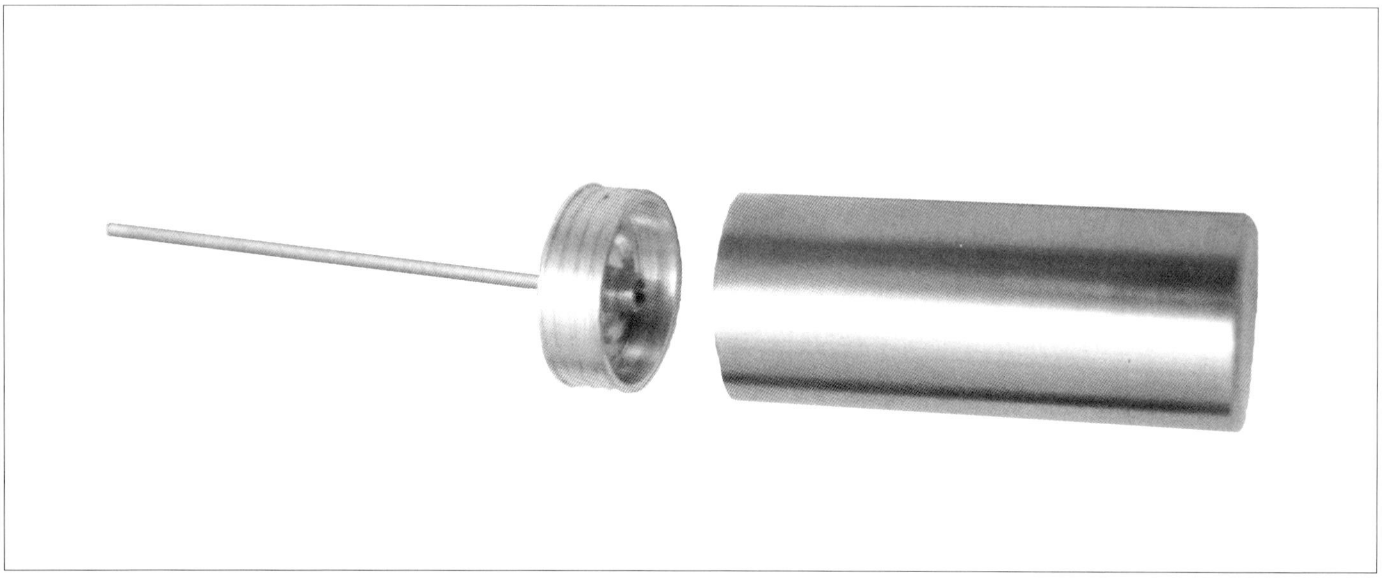

Photo 41: Displacer, plug and rod

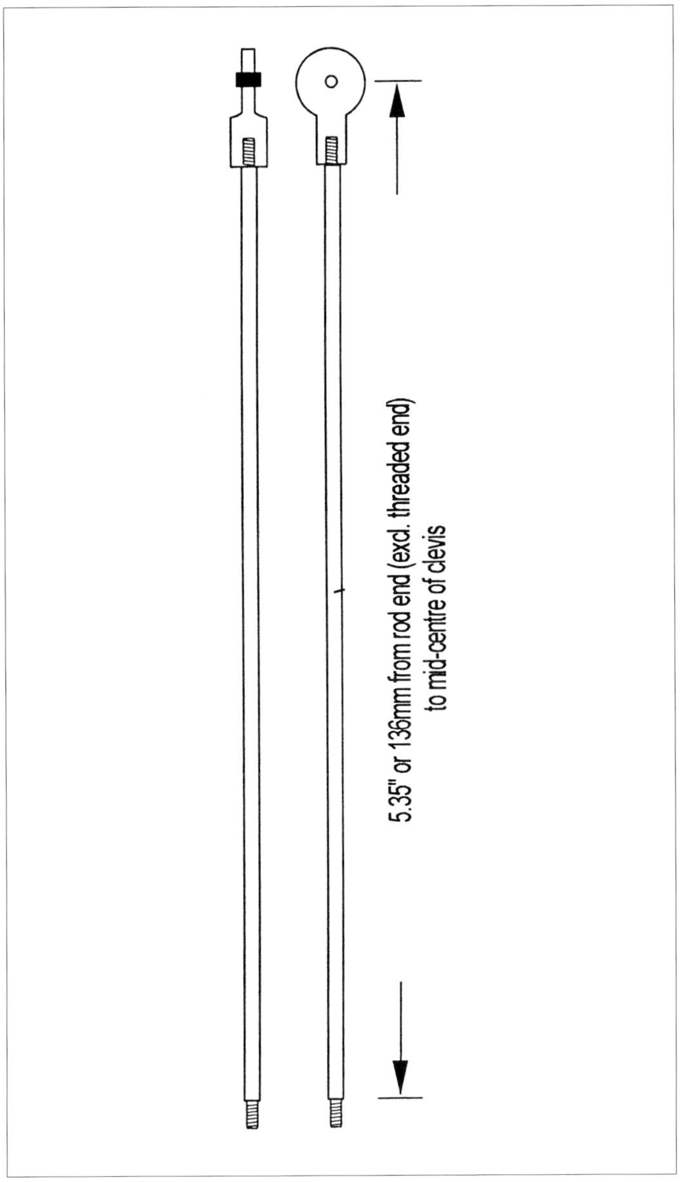

Fig. 10.24: Displacer rod (Approx. actual size)

18 DISPLACER CRANK, CRANKPIN, CON-ROD BIG-END AND CON-ROD

DISPLACER CRANK

(NOTE: The displacer crank of the Glasgow model is a relatively new cast item made in more recent years – and frankly it jars. The method used to secure the crank to the crankshaft is also unlike any other crank either in the Edinburgh model or the also relatively new power crank of the Glasgow model. The hump-like closure is clearly an unusual shape although its shape helps in two ways – easy to adjust the angle and less damage to the crankshaft. The side elevation shown in Figure 10.26 is a good copy as the shape was quite clear from a close-up image).

The front-end shape, length and width of the crank are shown in Figure 10.27. The stroke of the displacer as given to the author is 62mm ($2\frac{7}{16}$+" = 2.44"), i.e. the distance between the centres of the crank holes is $1\frac{7}{32}$" = 1.22" or 31mm. The remaining specifications are extrapolated from a close-up front image of

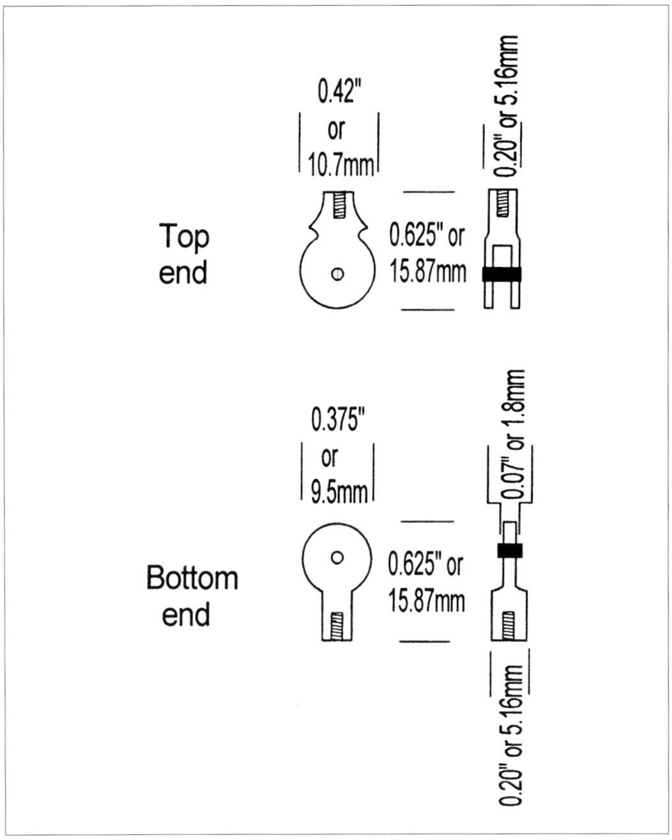

Fig. 10.25: Clevis device (Approx. actual size)

the component and compare quite reasonably with the displacer crank of the Edinburgh model and the power crank of the Glasgow model.

For those machinists who prefer the original crank as also shown in the Edinburgh model the following are the (similar) dimensions: Length: $1\frac{49}{64}$" = 1.77" (45mm), width at top end $\frac{7}{16}$–" = 0.43" (11mm), width at narrow end $\frac{5}{16}$" = 0.31" (8mm), thickness at top end $\frac{7}{32}$+" = 0.228" (5.8mm) and thickness at narrow end $\frac{11}{64}$–" = 0.169" (4.3mm)}.

The cast crank which is at present on the Glasgow model has a split end on the top part that tightens the crank into position (as against the grub screw type of the Edinburgh model).

CRANKPIN

The crankpin (Fig. 10.28) is machined from $\frac{5}{32}$–" = 0.151" (3.85mm) mild steel $\frac{13}{16}$+" = 0.826" (21mm) long, reduced at one end to $\frac{3}{32}$+" = 0.944" (or 2.4mm) for a length of $\frac{9}{32}$+" = 0.291" (7.4mm). This is the end that goes into the displacer con-rod big end, held in place with a pin. In fact the crankpin is cross-drilled to take a $\frac{3}{64}$" = 0.046" or 1.2mm pin to secure it in place once the big-end is inserted into the crankpin.

The other end of the crankpin is threaded $\frac{1}{8}$" = 0.125" (3.2mm) for a length of $\frac{7}{32}$–" = 0.212" (5.4mm). This is the end that will be threaded into the crank. (Fig. 10.27).

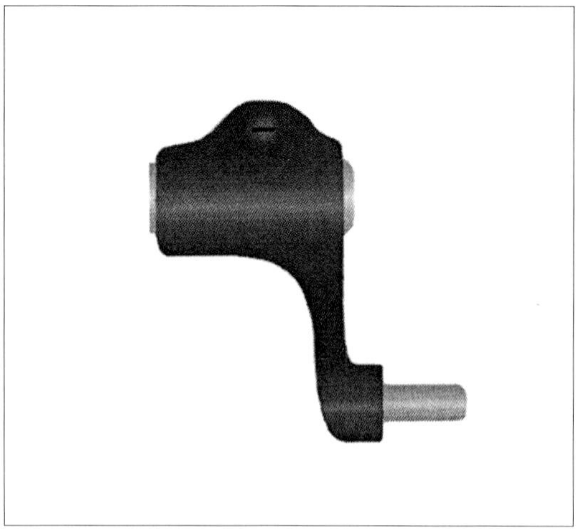

Fig. 10.26: Replacement cast displacer crank

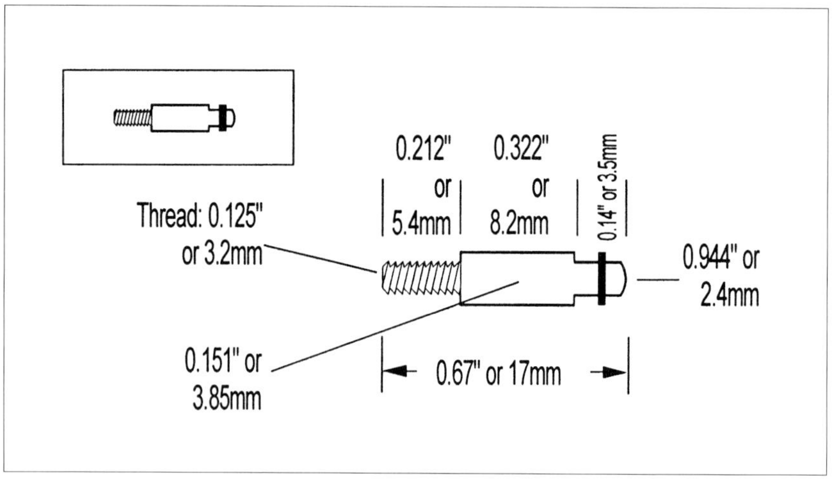

Fig. 10.28: 200% enlarged crankpin with specifications of an actual size one (Inset approx. size of the actual pin)

DISPLACER CON-ROD BIG-END

The big-end or the swivel end of the con-rod is machined from brass 3⁄16" = 0.1875" (4.8 or 5mm) thick, shaped as in Figure 10.29, with the round top part having a circumference of 3⁄8" = 0.375" (9.5 or 10mm). The lower end, which takes the threaded end of the con-rod, is machined 9⁄64" = 0.14" or 3.5mm square, drilled and tapped 8 BA or 2.5mm. The round end is drilled 3⁄32" = 0.0937" or 2.4mm to take the crankpin.

DISPLACER CON-ROD

The displacer con-rod is cut from 3.5 or 3.2mm (1⁄8") mild steel rod, approximately 31⁄32" = 3.03" (77mm) long, threaded one end 6 BA or 3mm (clevis end) and the other end 8 BA or 2.5mm (con-rod end). The actual length between the two components at the end (clevis and big-end) is 25⁄8" – 2.63" (67mm), the threaded ends being 13⁄64" = 0.20" (5mm), Figure 10.30. When assembled the distance between the centres of the clevis and the big-end is 317⁄32" – 3.54" or 90mm.

(Note: as explained earlier the use of L/H and R/H taps and dies facilitate the minor adjustments that may be necessary to obtain the correct stroke.)

19 FULL SCALE ASSEMBLY AND TESTING – STAGE 5

Refer to Figures 10.31a and 10.31b

The displacer unit is assembled separately from the engine frame, with the following sequence:

1 Displacer rod inserted into displacer plug with a smear of *Super Steel Araldite* or similar glue;
2 Displacer cylinder is inserted into the cooling jacket with the top end of the jacket in line with the top of the displacer cylinder sealing flange. The jacket and cylinder are silver soldered or brazed.
3 Displacer is inserted into the displacer cylinder.
4 The displacer rod/gland is screwed into the displacer cylinder head with a thin "O" ring and a smear of a suitable sealing liquid.
5 The displacer cylinder sealing flange is screwed to the cylinder (keeping in mind that the pressure pipe outlet is just over a bracket and opposite the outlet pipe of the cooling jacket).
6 *(At this stage the power cylinder base is unscrewed from the engine platform but retained in place possibly even by using temporary glue such as 'Blu Tack' by Bostik.)*

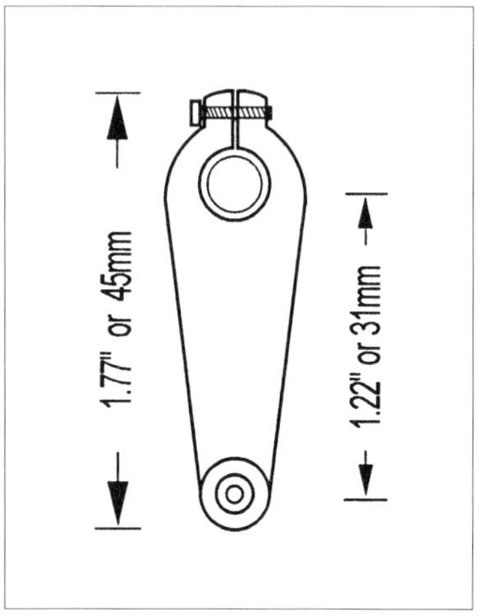

Fig. 10.27: Specifications of the displacer crank

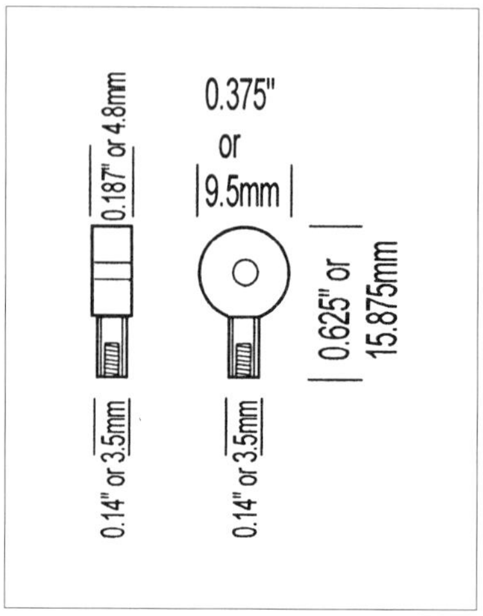

Fig. 10.29: Displacer con-rod big-end (Approx. size)

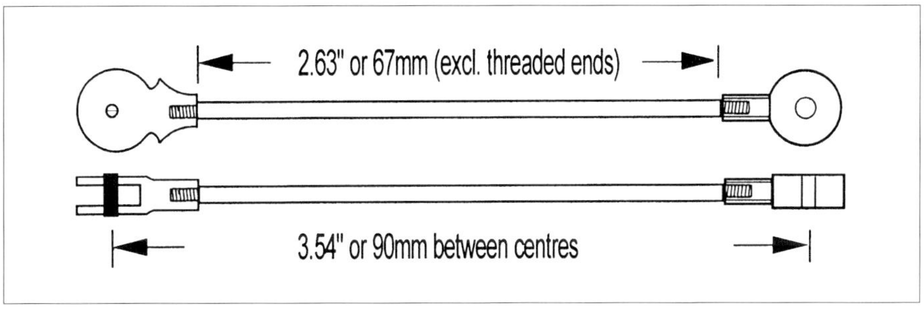

Fig. 10.30: Displacer con-rod (Actual size)

7 The whole displacer unit is then slid into the engine platform from above and placed such that the pressure pipe outlet is in line with the pressure pipe from the power cylinder.

8 Before screwing the displacer unit to the engine base and before soldering or gluing the pipe connections, the displacer con-rod from the crank is fitted to the displacer rod through the clevis device to ensure alignment also from this end.

9 When the vertical and horizontal alignments are confirmed, the cooling jacket brackets are screwed to the engine platform. It is suggested then only one bracket is screwed on and with one screw, until the pipe connection with the power cylinder is finished.

10 (The power cylinder is once more placed in position, the pressure pipe with the socket and sleeve slid into the sleeve/ connector at the displacer end with Araldite, (or Super Adhesive, Super Steel etc.), and then screwed down to the engine base).

11 The power and displacer cranks are set at 90° with the power crank at 12 and the displacer crank at 3 o'clock so that the flywheel will turn in the rear direction (if seen from the displacer side the flywheel will turn clockwise). (Photo 42).

BURNER and COOLING ARRANGEMENT

As with the Edinburgh model burners can be of a simple kind such as a spirit/wick container (see Figures 10.31a and 10.31b) or a more elaborate gas burner. The Glasgow model, having a thicker displacer cylinder base, will take a little longer to start until it heats up properly. On the other hand when connected to a water tank, the cooling arrangement ensures that the engine runs on for much longer than the Edinburgh model. For this reason two different burners have been designed to give a longer run for the Glasgow model.

A suitable Spirit Burner

This burner (Figure 10.32) has a long tank and nozzle with a fairly thick wick. The tank is cut from brass or copper tube with a thick wall, not less than 1⁄16" or 1.5mm, preferably 3⁄32" or 2.5mm or over. The tank is sealed at the bottom with a 2mm brass/copper disc. The disc and the bottom wall of the tube are both chamfered before soldering, and then filed or faced on the lathe. (Photo 43).

The top disc, at least 2mm thick, preferably thicker, is cut slightly larger in diameter by about 1⁄16" = 0.625" or 1.5mm all round, drilled in two places, as far apart as possible to take the nozzle and the filler cap. The hole for the nozzle is drilled 5⁄16" or 8mm, and slightly chamfered, while that of the filler cap is drilled and tapped ¼" x 24 or 26 t.p.i. or 6mm x 0.75mm. The top edge of the tank is chamfered, placed inverted on the disc and soldered.

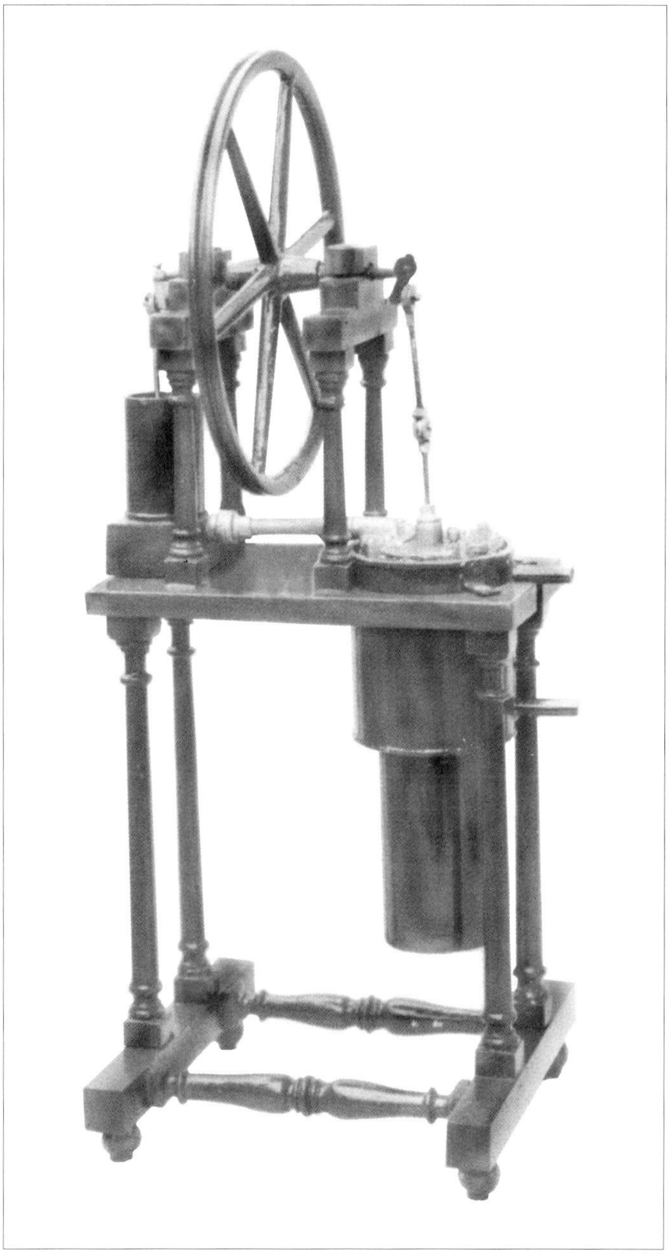

Photo 42: Full assembly of the replica Glasgow model

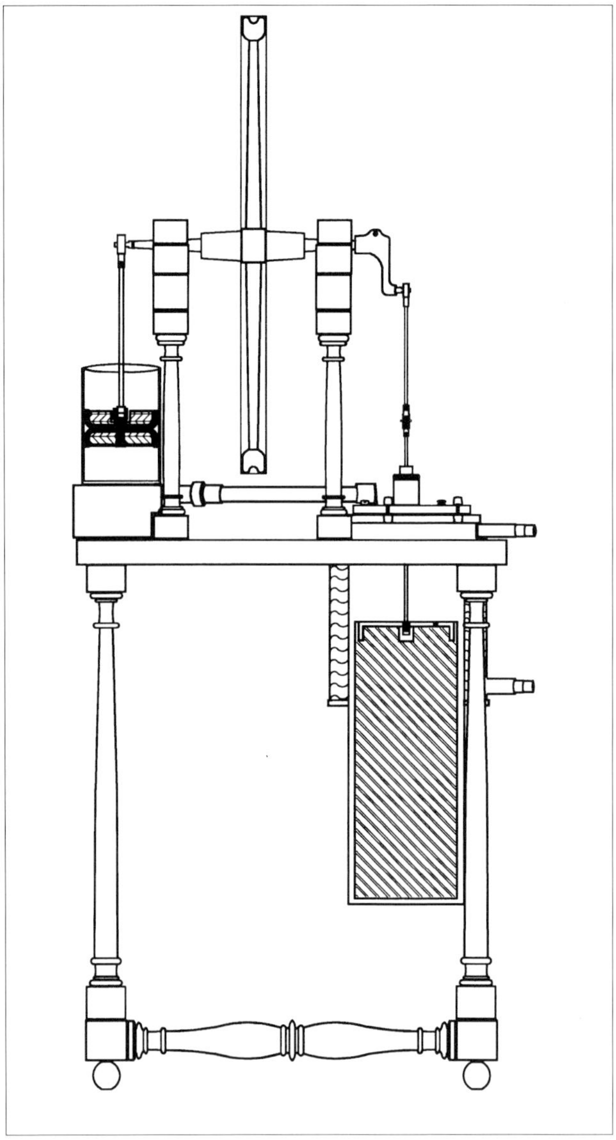

Fig. 9.31a: Sub-assembly - Stage 5 Displacer unit on engine platform connected to power cylinder (Reduced scale)

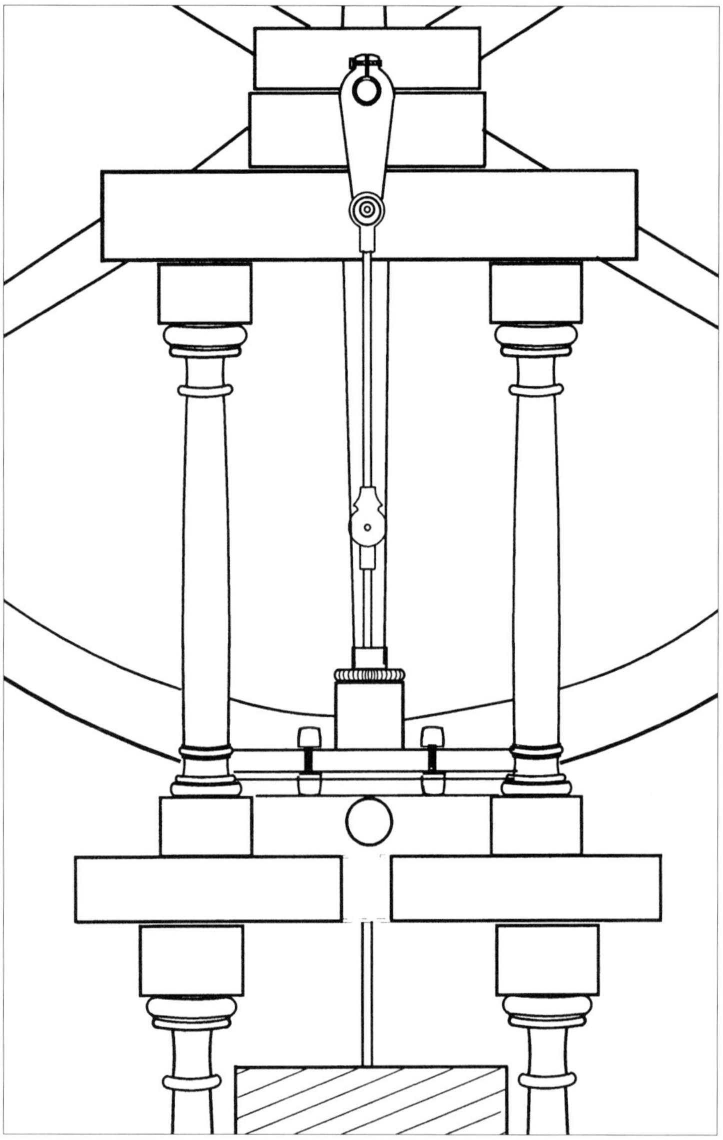

Fig. 10.31b: Full assembly - Stage 5 Detail from a side elevation of the assembled model with super-imposed crank, displacer con-rod and the position of the displacer (shaded)

Photo 43: Spirit burner suitable for long runs

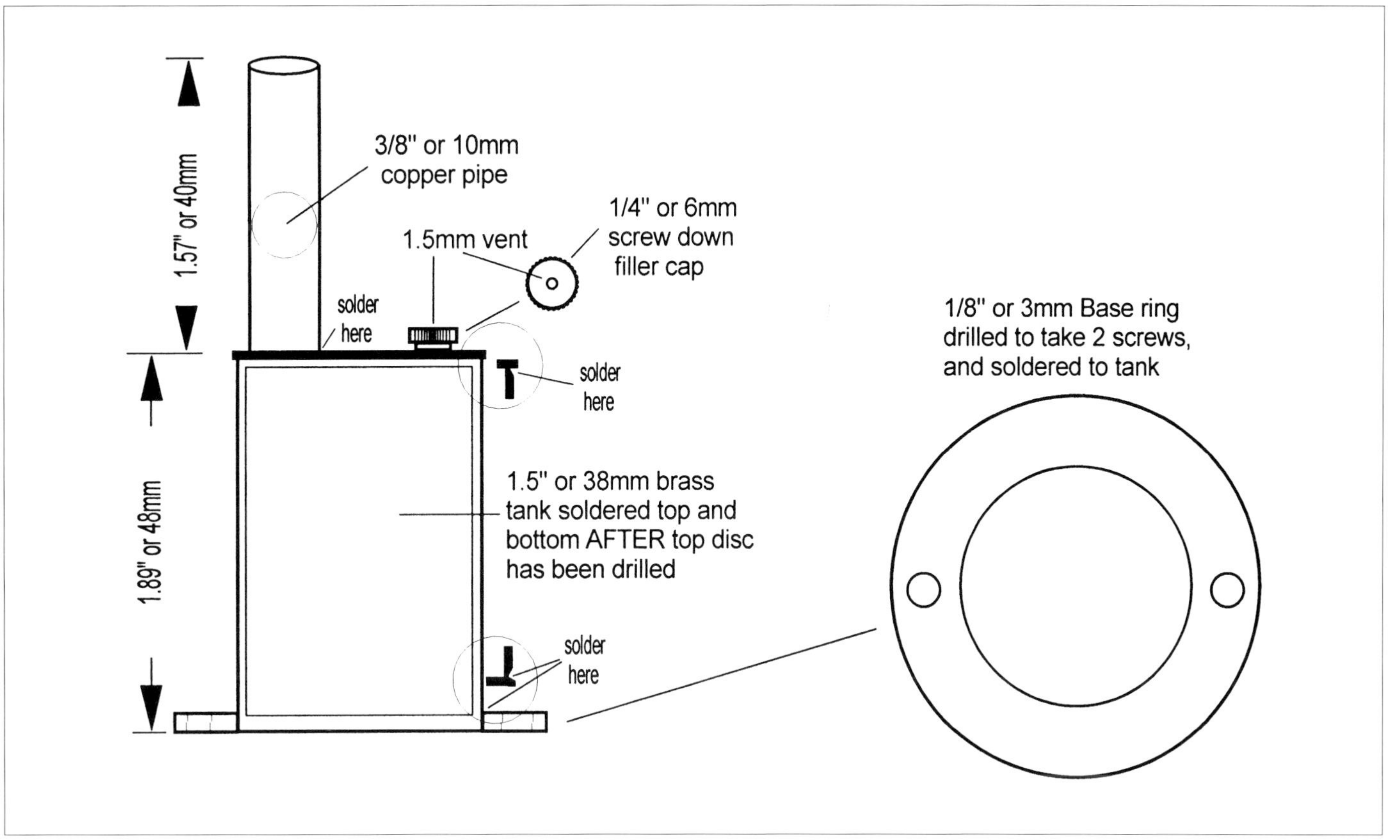

Fig. 10.32: Spirit burner suitable for a long-running Glasgow model

A ⅛"or 3mm brass disc, washer type, 2" or 50mm O.D., 1½" or 38mm I.D. is prepared as a base for the tank. This disc is drilled in two opposite places to take screws in case there is need to bolt it to a wooden base, then the bottom of the tank and the underside inner edge of the disc are filed and chamfered for soldering from underneath.

The next operation is fitting the nozzle. The wick, either round or flat (as in Photo 44) is first inserted into the copper tube. The wick is packed tight into the nozzle, and should be long enough to reach the bottom with a little to spare, while at the top the wick is first left with ½" or 12mm protruding. The centre of the top half of the tube is squashed between two metal or hardwood bars of about 1" diameter and then the top end is flared in the opposite direction by gently squashing it in a vice to give an oblong finish. The nozzle is inserted into the tank by about ¼" or 6mm and soldered.

The filler cap is machined from ½" or 12mm O.D. brass bar, ½" long, reduced and threaded the same as the hole in the top disc of the tank for a length of ¼". The top edge is rounded and either knurled or filed to give a grip, then drilled ¹⁄₁₆" or 1.5mm in the centre right through. This is an important vent to allow gas generated by the heat to escape. The filler cap is inserted into the tank with a fibre washer ('O' rings are to be avoided as they tend to melt if the tank top gets too hot).

The wick is cut to about ¼" above the flared top, but if during a burn test the flame is too high, the wick is cut down, but not too far since the wick will burn in time and enough length must be retained to pull up more wick length through the nozzle, bearing in mind that the wick is squashed in the tube and therefore slightly hard to extract.

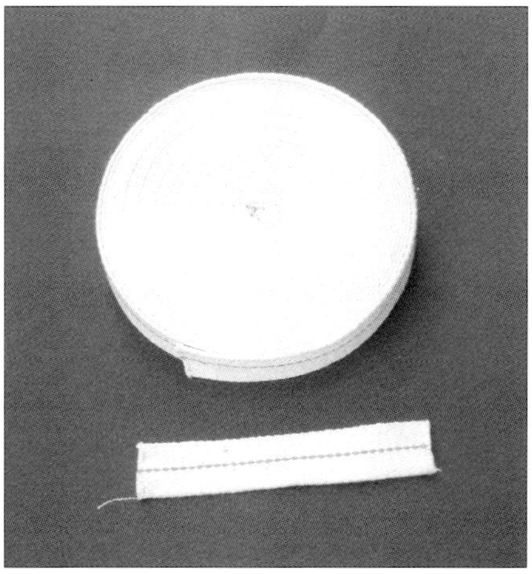

Photo 44: Type of wick used for the spirit burner

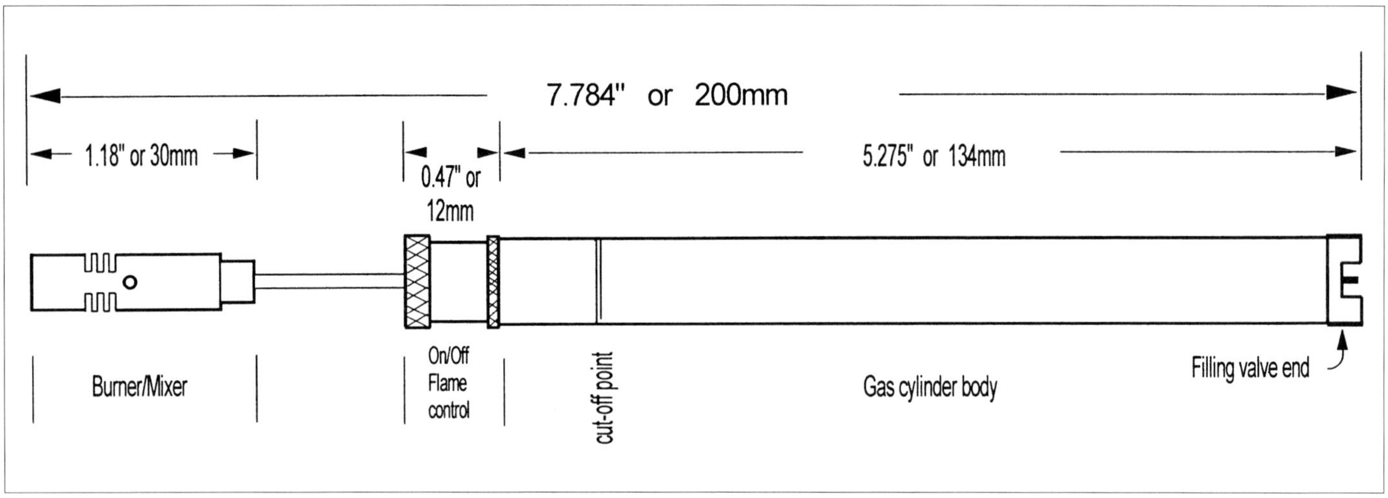

Fig. 10.33: Profile and specifications of gas pencil burner

A re-constructed gas pencil burner

This pencil burner had been adapted for a number of small Stirling engines and was found to be very efficient, heating the bottom end of the displacer cylinder quite quickly and capable of maintaining a run of about 15 minutes duration. These pencil burners are readily available from stands in engineering exhibitions and from hardware stores. (Figure 10.33)

The burner can be changed into a vertical flame spreader with some basic machining. ***The burner is first completely emptied of all gas.*** With a fine hacksaw the tube is cut at a distance of ⅝" = 0.625" or 15mm from the knurled flame adjuster or 3⁹⁄₆₄" = 3.14" or 80mm from the top end. The cut is made with a fine hacksaw cutting round the burner body without going too deep or penetrating the internal foam. The foam is cut with a sharp knife.

An oblong aluminium block 1" or 25mm square and 1¹¹⁄₃₂" = 1.34" (34mm) is drilled from the top and from one end to meet slightly off-centre as in Figure 10.34. The top part of the burner (vertical) is first inserted into the hole in the surface of the block with a smear of *Super Steel epoxy (by Plastic Padding)*. When the epoxy has hardened the cavity is tightly filled with a similar type of foam as is seen in the burner body, and then the horizontal burner body is also inserted with the same epoxy. The burner is left overnight for the epoxy to get hold on the various surfaces. (Photo 45)

The burner is tried with only a partial filling of gas and left standing for some hours to check for any leakage. Apart from proprietary brands of gas leak detectors, a few drops of soapy water round the two bores and tube insertions will immediately show signs of leakage. Later the gas cylinder can be filled as per instructions given with the original burner. (At first the flame will be spluttering and not easy to control, but when the first gas has evaporated, a steady blue flame will give the right amount of heat to the Stirling model.)

If it is intended that this gas burner to be a permanent fixture, a couple of small clips screwed to a wooden base ensure that the burner will neither move nor accidentally tip over.

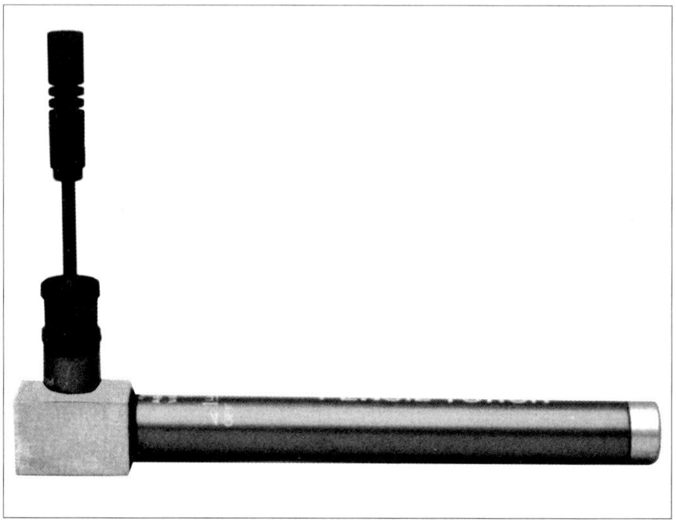

Photo 45: Re-constructed gas pencil burner with vertical flame thrower

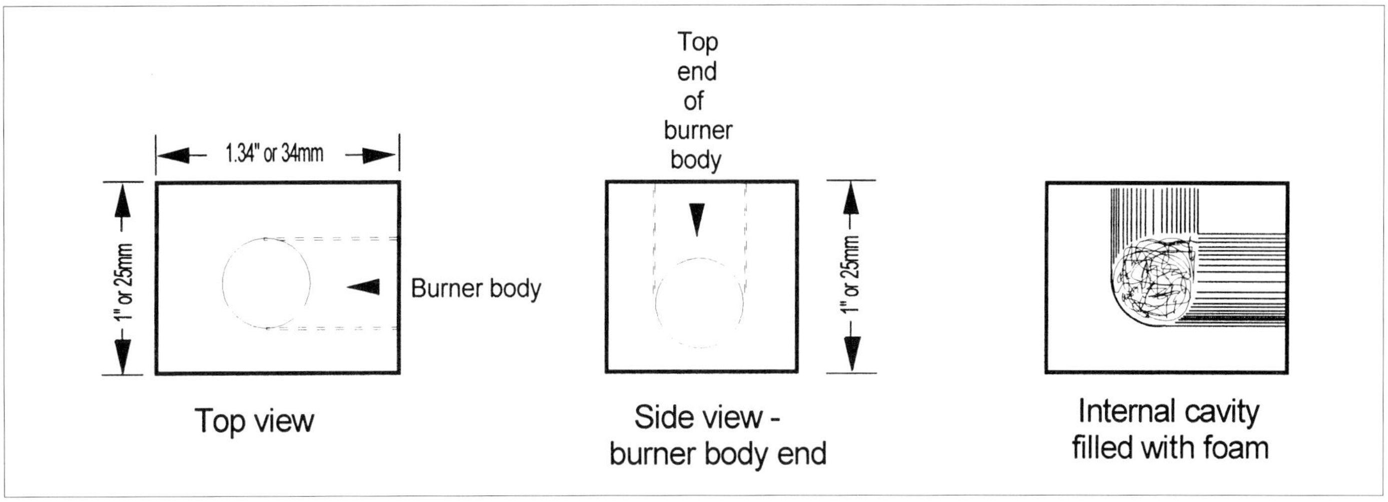

Fig. 10.34: Aluminium base for re-constructed burner

The Cooling Tank

A type of cooling tank, made from a long, even narrow, tin or aluminium containers, with an inlet and outlet pipes connected to the engine will maintain heat separation for a long time. A most suitable cooling tank can be made from aluminium square profiles used in the construction industry, normally about 2" or 50mm square, the bottom sealed with an aluminium plug and base, and drilled to take two short pieces of brass or aluminium pipe and glued with *Super Steel, Super Epoxy* or similar strong adhesive. Another alternative is an aluminium round tube, same diameter as above, finished in the same way. Both these cooling tanks should have a ¼ turn stop cock (plastic will do) near the bottom for emptying. (Photo 46)

Running the Glasgow Model

As explained above, this model having a thick base will take a little longer than the Edinburgh model to start, but on the other hand with the water cooling arrangement it will run for a much longer duration. The displacer cylinder should be allowed to heat up for a few minutes before any attempt is made to turn the flywheel.

These models were not made for fast revolutions, and in fact much of the pleasure and interest is lost if the performance of the con-rods, cranks and the flywheel is blurred, apart from the possibility of damage to the frail wooden structure.

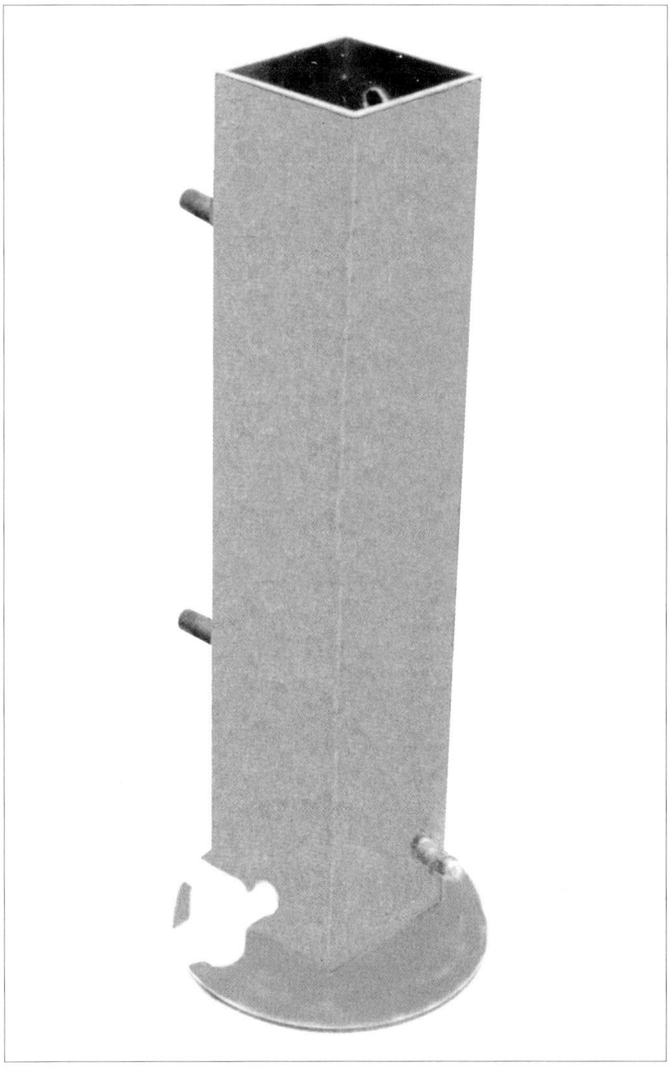

Photo 46: Cooling tank used to run the Glasgow model replica

Chapter II
Fault Finding and Trouble Shooting

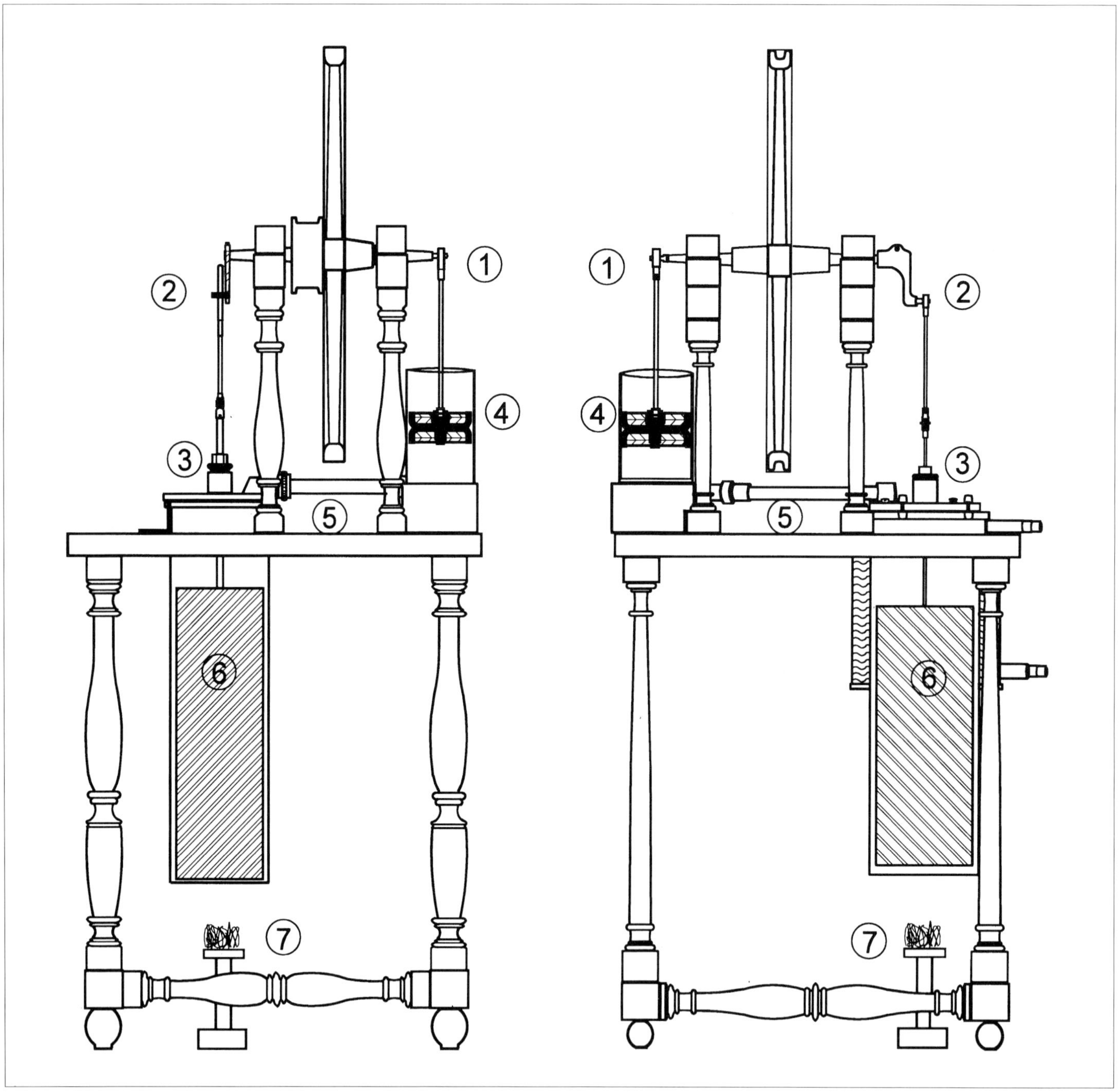

1 Friction in bearings - too tight

2 Friction in cranks, con-rods/wrist joints, clevis

3 Escaping air pressure from gland or seating, rods too narrow; also rod fitting too tight

4 Piston too tight; piston too loose: (should be gas-tight but with minimum friction)

5 Escaping air pressure from connecting pipe and/or any socket fitting

6 Displacer not air-tight; displacer too heavy; too much gap at top and bottom of the stroke (ideally not more than 1⁄16" or 1.5mm)

7 Insufficient heating (too high heating will cause the engine to slow and eventually stop)

Appendix I
Specifications of Robert Stirling's Edinburgh Model

NOTE: Those dimensions marked with an asterisk (*) were given to the author as at source. These dimensions were measured. The remaining specifications were calculated on the basis of extrapolation or calculation from photographs – mostly close-up – and are as close as possible (and feasible) to the original.

Some fraction inch measurements have either (+) or (-); the difference is slight, sometimes by less than $\frac{1}{128}$" which does not affect either the layout, design or function of the model.

Component	Dimension	Imperial (inches)		Metric
		fractions	decimal	mm
*Engine	total height	29		737
*Top engine platform	length	$12\frac{1}{2}$	12.5	317
	width	6		152
	thickness	$\frac{11}{16}$	0.687	17.46
Bore for displacer cylinder		$3\frac{5}{32}$	3.15	80
Bore for elliptical hole	length	$1\frac{19}{64}$+	1.30	33.06
	width	$1\frac{5}{32}$	1.153	29.29
*Top turned columns (4)	length	7		178
	capitals & pedestals	$\frac{27}{32}$+	0.848	21.55
*Lintels (2)	length	$5\frac{35}{64}$	5.546	140.88
	height	$\frac{31}{32}$	0.976	24.8
	width	$\frac{31}{32}$	0.976	24.8
*Crankshaft securing brackets	length	$2\frac{3}{64}$	2.04	51.80
	height	$\frac{51}{64}$+	0.803	20.40
	width	$\frac{29}{32}$+	0.905	23
*Bottom turned columns (4)	length	12		304.8
	capitals & pedestals	$1\frac{5}{64}$+	1.088	27.63
*Base cross-bars (2)	length	$11\frac{1}{2}$ -	11.49	292
	height	$1\frac{3}{32}$	1.09	27.7
	width	$1\frac{3}{32}$	1.102	28
*Balusters (horizontal turned pieces - 2)	length	$9\frac{3}{8}$	9.375	238
	square ends	1		25.4

*Flywheel:	O.D.		12		304.8
	rim thickness	$\frac{5}{8}$		0.625	15.875
	rim depth	$\frac{9}{16}$		0.562	14.28
*Pulley	maximum O.D.	$2\frac{13}{64}$		2.204	56
	external width	$\frac{31}{32}$		0.97	24.63
	internal I.D.	$1\frac{57}{64}$		1.89	48
	inside pulley width	$\frac{39}{64}$+		0.612	15.54
*Flywheel fillet	total length	$\frac{61}{64}$		0.954	24.24
	O.D. near flywheel	$\frac{15}{16}$		0.944	24
	…..tapering to	$\frac{45}{64}$+		0.708	18
*Crankshaft	length	$6\frac{19}{64}$		6.3	160
	O.D. at pulley and fillet ends	$\frac{3}{8}$		0.375	9.5
	…..tapering to	$\frac{7}{32}$		0.228	5.8
POWER UNIT					
*Power cylinder	external length	$3\frac{1}{8}$+		3.137	79.7
	internal length	$3\frac{1}{64}$		3.02	76.7
	O.D.	$2\frac{1}{16}$-		2.06	52.32
	I.D.	$1\frac{61}{64}$		1.95	49.62
	flange	$2\frac{11}{64}$		2.18	55.42
*Power cylinder base					
	width from front	$2\frac{17}{64}$		2.267	57.6
	length from side	$2\frac{53}{64}$		2.834	72
	height	$1\frac{27}{64}$		1.418	36.04
Power piston	length		1		25.4
*Power con-rod	mid-centre of big-end to mid-centre of small-end	$4\frac{29}{64}$-		4.45	113
*Power Crank:	total length	$1\frac{25}{64}$		1.393	35.40
	length between centres	$\frac{21}{32}$		0.657	16.705
	thickness	$\frac{13}{64}$		0.20	5.08
	…..tapering to	$\frac{5}{32}$+		0.163	4.15
Crankpin	length	$\frac{11}{64}$		0.69	17.64
	O.D. of pin	$\frac{5}{32}$		0.15	3.85
Power con-rod big-end	total length		1		25.4
	round part	$\frac{31}{64}$		0.48	12.2
	thickness	$\frac{3}{16}$		0.178	4.57
	hexagon part	$\frac{25}{64}$		0.39	10
*Power piston	stroke	$1\frac{5}{16}$		1.31	33.41

DISPLACER UNIT				
.*Displacer cylinder	total length	$11\frac{1}{8}$	11.126	282.6
	internal length	$11\frac{1}{16}$	11.063	281
	O.D.	$2\frac{31}{32}$	2.97	75.52
	I.D.	$2\frac{7}{8}$	2.87	73
	flange	$3\frac{9}{32}$	3.285	83.46
Sealing top/flange	O.D.	$3\frac{3}{8}$	3.383	85.94
	total thickness	$\frac{15}{64}$	0.236	6
	part thickness	$\frac{5}{32}$	0.156	3.98
*Displacer	length	$8\frac{1}{2}$	8.5	216
	O.D.	$2\frac{53}{64}$	2.834	72
	I.D.	$2\frac{3}{4}$	2.756	70
Displacer rod	diameter			3.2 or 3.5
	length	$5\frac{1}{32}$	5.04	128
Displacer con-rod	length	$4\frac{1}{2}$	4.508	114.5
	length between centres of clevis and crankpin	$3\frac{27}{64}$	3.425	87
*Displacer rod guide bush	total length of two parts	$1\frac{5}{8}$	1.632	41.47
	*net length	$1\frac{7}{64}$	1.115	28.33
	*O.D.	$\frac{47}{64}$	0.728	18.5
*Displacer crank	total length	$1\frac{43}{64}$	1.677	42.59
	length between centres	$1\frac{11}{64}$	1.176	29.88
	thickness	$\frac{13}{64}$	0.206	5.25
	width - top at crankshaft level	$\frac{15}{32}$	0.466	11.85
	width - bottom at c/pin level	$\frac{11}{32}$	0.354	9
*Crankpin	diameter	$\frac{9}{64}$+	0.151	3.85
	length	$\frac{53}{64}$	0.875	21
*Displacer:	stroke	$2\frac{11}{32}$	2.35	59.7 - 60
*Yoke	total external length	$2\frac{5}{64}$	2.086	53
	net length	$1\frac{5}{8}$+	1.634	41.52
	yoke width	$\frac{3}{8}$	0.374	9.5
	yoke thickness	$\frac{9}{64}$	0.137	3.5
	slot width	$\frac{5}{32}$-	0.152	3.88
	slot length	$1\frac{3}{8}$	1.372	34.87
	securing nut:	$\frac{3}{16}$	0.187	4.76
	washer O.D.	$\frac{3}{8}$	0.375	9.6

Appendix 2
Specifications of Robert Stirling's Glasgow Model

(SEE NOTE IN APPENDIX 1)

Component	Dimension	Imperial (inches) fractions		Imperial (inches) decimal	Metric mm
*Engine	total height		29		737
*Top engine platform	length	$11\frac{31}{32}$		11.97	304
	width	$6\frac{9}{64}$		6.14	156
	thickness	$\frac{45}{64}$		0.708	18
	bore for cooling jacket	$4\frac{41}{64}$		4.645	118
*Top turned columns (4)	length	$6\frac{3}{32}$		6.10	155
	capitals & pedestals	$\frac{63}{64}$		0.98	25
*Bottom turned columns (4)	length	$12\frac{1}{8}$		12.125	308
	capitals & pedestals	$1\frac{3}{32}$		1.10	28
*Base cross-bars (2)	length	$11\frac{7}{32}$		11.22	285
	height	$1\frac{1}{16}$		1.063	27
	width	$1\frac{3}{32}$		1.10	28
*Balusters (horizontal turned pieces - 2)	net length	$8\frac{55}{64}$		8.86	225
	square ends	$1\frac{1}{16}$		1.063	27
*Lintels (2)	length	$5\frac{43}{64}$		5.67	144
	height	$\frac{29}{32}$		0.90	23
	width	$\frac{63}{64}$		0.984	25
*Crankshaft-bearing base (2)	length	$2\frac{9}{16}$		2.56	65
	height	$\frac{3}{4}$-		0.748	19
	width	$\frac{63}{64}$		0.984	25
*Crankshaft-bearing cap (2)	length	$2\frac{23}{64}$		2.36	60
	height	$\frac{5}{8}$+		0.63	16
	width	$\frac{15}{16}$		0.94	24

*Flywheel	O.D.	$12\frac{9}{32}$		12.283	312
	rim width	$\frac{5}{8}$+		0.63	16
	rim depth	$\frac{33}{64}$		0.511	13
*Flywheel fillet	total length	$1\frac{1}{64}$		1.063	27
	O.D. near flywheel	$\frac{13}{16}$		0.811	20.6
	…..tapering to	$\frac{45}{64}$		0.708	18
*Crankshaft	length	$7\frac{51}{64}$		7.795	198
	O.D.	$\frac{3}{8}$		0.375	9.5
	…..tapering to	$\frac{7}{32}$		0.228	5.8
POWER UNIT					
*Power cylinder "	external length	$3\frac{9}{64}$+		3.15	80
	internal length	$3\frac{1}{32}$		3.03	77
	O.D.	$2\frac{5}{32}$		2.165	55
	I.D.	$2\frac{3}{64}$		2.047	52
	flange	$2\frac{9}{32}$		2.283	58
*Power cylinder platform	width from front	$2\frac{31}{64}$		2.48	63
	length from side	$2\frac{9}{16}$		2.56	65
	height:	$1\frac{13}{32}$		1.417	36
*Power piston	length		1		25.4
*Power con-rod	mid-centre of big-end crankpin to mid-centre of small-end	$4\frac{31}{64}$		4.48	114
*Power Crank	total length:	$1\frac{11}{64}$		1.18	30
	length between centres:	$\frac{53}{64}$		0.825	21
	thickness	$\frac{13}{64}$		0.20	5.08
	…..tapering to	$\frac{5}{32}$		0.163	4.15
	width at crankshaft level	$\frac{7}{16}$		0.437	11.11
	width at crankpin level	$\frac{21}{64}$		0.326	8.3
Power piston	stroke	$1\frac{21}{32}$-		1.653	42
DISPLACER UNIT					
*Water cooling jacket	length:	$5\frac{5}{32}$		5.157	131
	O.D.	$4\frac{21}{64}$		4.33	110
	I.D	$4\frac{1}{4}$		4.25	108
*Displacer cylinder	external length total	$10\frac{1}{2}$+		10.51	267
	internal length:	$10\frac{25}{64}$		10.39	264
	O.D.	$3\frac{7}{32}$+		3.22	82
	I.D.	$2\frac{63}{64}$+		2.99	76
	flange	$3\frac{25}{32}$		3.78	96

Sealing top/flange	O.D.	$3\frac{25}{32}$	3.78	96
	total thickness:	$\frac{15}{64}$	0.234	6
*Displacer	length	$7\frac{33}{64}$	7.52	191
	O.D.	$2\frac{7}{8}$	2.87	73
	I.D.	$2\frac{51}{64}$	2.8	71
Displacer rod	diameter:			3.2 or 3.5
	length:	$5\frac{3}{16}+$	5.196	132
Displacer con-rod	length between centres of clevis and big-end:	$3\frac{17}{32}$	3.54	90
*Displacer rod guide bush	total length when screwed	$1\frac{19}{64}$	1.3	33
	O.D.	$\frac{45}{64}+$	0.708	18
*Displacer crank	total length:	$1\frac{49}{64}$	1.77	45
	length between centres	$1\frac{7}{32}$	1.22	31
	thickness	$\frac{7}{32}$	0.218	5.5
	width top at crankshaft level	$\frac{7}{16}-$	0.43	11
	width bottom at crankpin level	$\frac{3}{16}$	0.31	8
*Displacer:	stroke	$2\frac{7}{16}$	2.44	62

Appendix 3
Conversion Tables

TABLE 1: Inches (fractions and decimal) to Millimetres

inches		mm	inches		mm	inches		mm
1/64	0.0156	0.3969	11/32	0.3437	8.7312	43/64	0.6718	17.066
1/32	0.0312	0.7938	23/64	0.3593	9.1281	11/16	0.6875	17.462
3/64	0.0468	1.1906	3/8	0.375	9.5250	45/64	0.7031	17.859
1/16	0.0625	1.5875	25/64	0.3906	9.9219	23/32	0.7187	18.256
5/64	0.0781	1.9844	13/32	0.4062	10.3190	47/64	0.7343	18.653
3/32	0.0937	2.3812	27/64	0.4218	10.7160	3/4	0.75	19.05
7/64	0.1093	2.7781	7/16	0.4375	11.1120	49/64	0.7656	19.447
1/8	0.125	3.175	29/64	0.4531	11.5090	25/32	0.7812	19.844
9/64	0.1406	3.5719	15/32	0.4687	11.9060	51/64	0.7968	20.241
5/32	0.1562	3.9688	31/64	0.4843	12.3030	13/16	0.8125	20.638
11/64	0.1718	4.3656	1/2	0.5	12.70	53/64	0.8281	21.034
3/16	0.1875	4.7625	33/64	0.5156	13.097	27/32	0.8437	21.431
13/64	0.2031	5.1594	17/32	0.5312	13.494	55/64	0.8593	21.828
7/32	0.2187	5.5562	35/64	0.5468	13.891	7/8	0.875	22.225
15/64	0.2343	5.9531	9/16	0.5625	14.288	57/64	0.8906	22.622
1/4	0.25	6.350	37/64	0.5781	14.684	29/32	0.9062	23.019
17/64	0.2656	6.7469	19/32	0.5937	15.081	59/64	0.9218	23.416
9/32	0.2812	7.1438	39/64	0.6093	15.478	15/16	0.9375	23.812
19/64	0.2968	7.5406	5/8	0.625	15.875	61/64	0.9531	24.207
5/16	0.3125	7.9375	41/64	0.6406	16.272	31/32	0.9687	24.606
21/64	0.3281	8.3344	21/32	0.6562	16.669	63/64	0.9843	25.003
						1	1.0000	25.400

1meter = 39.370113 inches; m/m to inches x 0.03937
1 inch = 25.399978m/m; inches to m/m x 25.4

TABLE 2: Millimetres to Inches (decimal)

mm inches	mm inches	mm inches
1 = 0.0394	34 = 1.3386	67 = 2.6378
2 = 0.0787	35 = 1.3780	68 = 2.6772
3 = 0.1181	36 = 1.4173	69 = 2.7165
4 = 0.1575	37 = 1.4567	70 = 2.7559
5 = 0.1969	38 = 1.4961	71 = 2.7953
6 = 0.2362	39 = 1.5354	72 = 2.8346
7 = 0.2756	40 = 1.5748	73 = 2.8740
8 = 0.3150	41 = 1.6142	74 = 2.9134
9 = 0.3543	42 = 1.6536	75 = 2.9528
10 = 0.3937	43 = 1.6929	76 = 2.9921
11 = 0.4331	44 = 1.7323	77 = 3.0315
12 = 0.4724	45 = 1.7716	78 = 3.0709
13 = 0.5118	46 = 1.8110	79 = 3.1102
14 = 0.5512	47 = 1.8504	80 = 3.1496
15 = 0.5906	48 = 1.8898	81 = 3.1890
16 = 0.6299	49 = 1.9291	82 = 3.2283
17 = 0.6693	50 = 1.9685	83 = 3.2677
18 = 0.7087	51 = 2.0079	84 = 3.3071
19 = 0.7480	52 = 2.0472	85 = 3.3465
20 = 0.7874	53 = 2.0866	86 = 3.3858
21 = 0.8268	54 = 2.1260	87 = 3.4252
22 = 0.8661	55 = 2.1654	88 = 3.4646
23 = 0.9055	56 = 2.2047	89 = 3.5039
24 = 0.9449	57 = 2.2441	90 = 3.5433
25 = 0.9843	58 = 2.2835	91 = 3.5827
26 = 1.0236	59 = 2.3228	92 = 3.6221
27 = 1.0630	60 = 2.3622	93 = 3.6614
28 = 1.1024	61 = 2.4016	94 = 3.7008
29 = 1.1417	62 = 2.4409	95 = 3.7402
30 = 1.1811	63 = 2.4803	96 = 3.7795
31 = 1.2205	64 = 2.5197	97 = 3.8189
32 = 1.2599	65 = 2.5591	98 = 3.8583
33 = 1.2992	66 = 2.5984	99 = 3.8976
		100 = 3.9370

APPENDIX 4
SELECTED BIBLIOGRAPHY

STIRLING AND HOT AIR ENGINES
by Roy Darlington and Keith Strong – published by The Crowood Press Ltd.

REV. ROBERT STIRLING, D.D.
by Robert Sier, published by L.A. Mair – a biography of the inventor of the Heat Economiser and the Stirling-Cycle Engine.

HOT AIR, CALORIC AND STIRLING ENGINES (Volume 1)
by Robert Sier, published by L.A. Mair.

AN INTRODUCTION TO STIRLING ENGINES
by Professor James R. Senft, published by Moriya Press.

BUILDING STIRLING ONE
by T. Warbrooke, published by Camden Miniature Steam Services.

THERMODYNAMICS AND GAS DYNAMICS OF THE STIRLING CYCLE MACHINE
by Dr. A.J. Organ, published by Cambridge University Press.

THE PHILIPS STIRLING ENGINE
by C.M. Hargreaves, published by Elsevier.

AN INTRODUCTION TO LOW TEMPERATURE DIFFERENTIAL STIRLING ENGINES
by Prof. James R. Senft, published by Moriya Press.

THE REGENERATOR AND THE STIRLING ENGINE
by Dr. A.J. Organ, published by M.E.P.

STIRLING CYCLE ENGINES
by Andy Ross, published by Solar Engines, Phoenix, Arizona, USA.

STIRLING ENGINES
by Prof. Graham Walker, published by Clarendon Press, Oxford.

STIRLING ENGINES
by G.T. Reader and C. Hooper, published by E. & F.N. Spon, London and New York.

RINGBOM STIRLING ENGINES
by Prof. James R. Senft, published by Oxford University Press, New York.

MAKING STIRLING ENGINES
by Andy Ross, published by Ross Experimental, Columbus, Ohio, USA,

(Readers, please be aware that not all the above books are currently in print)

ADDITIONAL READING:

'*Engineering in Miniature*' and '*Model Engineer*' in the U.K., '*Australian Model Engineering*' in Australia, and '*Live Steam*' in America often print articles by Stirling Engine enthusiasts and builders. They are worth reading and making models of the engines described. Newcomers to this hobby are recommended to look into back numbers of these magazines for such articles.

A number of historical and informative articles appeared in '*The Engineer*' in the 19th and early 20th centuries, but one particular and more recent study is a must for Stirling Engine modellers who wish to know more about the history of this engine. These articles were written by Dr. T. Finklestein, entitled "Air Engines", and appeared in the 1959 issues, Volume 207, Pages 492-497, 522-527, 720-723.

Appendix 5
The 'Stirling Engine Society'

The 'STIRLING ENGINE SOCIETY' was formed in January 1997 when a group of enthusiasts who meet regularly during the various national engineering exhibitions decided to put their comradeship and their common interest on a more formal basis.

The aim of the Society is to unite engineers, hobbyists, experimenters and enthusiasts involved with Stirling Cycle Engines into an international association dedicated to promoting the development of the Stirling Engine as a prime mover, as well as furthering understanding of Stirling Cycle machines. The Society has attracted quite a number of Members from various countries and universities.

The Society participates in major engineering and hobby exhibitions in England and regularly has its own stand on which a great number of various types of Stirling Engines are run for the benefit of visitors. Additionally some Members also take part and set up stands in regional and local exhibitions. Members altogether spend countless hours explaining the technology and principles of the Stirling-Cycle, as well as demonstrating the engines on stand to the members of the public who regularly flock to see these 'unusual' machines operating with just a small gas flame or a spirit burner.

The Society has taken part in a number of conferences and exhibitions in Europe and the USA, either as a Society or with individual member participation. This has further enhanced interest in the development of the Stirling-cycle engine as an excellent contender for environmentally-friendly motive power.

The Society also publishes a newsletter, "*Stirling News*", which gives the latest technological advances, contributions from many parts of the world, news of various exhibitions, construction articles, etc.. Some back numbers may still be available for new Members.

The Society organizes the enthusiasts in specialist interest groups so that information may be readily exchanged between those Members who indicate their preference.

Further details may be obtained on www.stirlingengines.org, or by post from: P.O. Box 5909, Chelmsford, CM1 2PG, U.K.

Photo 47: Some of the Founder/Members of the Stirling Engine Society. Front row (left to right): the author, Richard White, John Wilkinson Back row (left to right): Roy Darlington, Geoff Bartlett, Julian Wood